Lubrication Strategies and Tips
How to "Kick-start" any Lubrication Program

Published 2024 by River Publishers

River Publishers

Alsbjergvej 10, 9260 Gistrup, Denmark

www.riverpublishers.com

Distributed exclusively by Routledge

605 Third Avenue, New York, NY 10017, USA

4 Park Square, Milton Park, Abingdon, Oxon OX14 4RN

Lubrication Strategies and Tips / by Kenneth E. Bannister.

Routledge is an imprint of the Taylor & Francis Group, an informa business

ISBN 978-87-7004-624-4 (paperback)

ISBN 978-87-7004-626-8 (online)

ISBN 978-8-770-04625-1 (ebook master)

A Publication in the River Publishers Series in Rapids

Lubrication Strategies and Tips
How to "Kick-start" any Lubrication Program

Kenneth E. Bannister

Canada

NEW YORK AND LONDON

Contents

Preface

Welcome to the world of lubrication—a realm where adages, myths, and misconceptions have woven themselves into the fabric of our understanding. As we embark on this journey, it's crucial to recognize the origins of these beliefs, rooted in the early 20th century agrarian landscape of North America. Back then, lubrication decisions were simpler, driven by robust machinery and limited choices in lubricants.

Fast forward to today, and we find ourselves in a vastly different landscape. Our machinery is sophisticated, with finer tolerances and intricate designs. Lubrication isn't just an afterthought; it's an integral part of the machine's performance and longevity. Yet, despite these advancements, myths and misconceptions still persist.

One such myth is the belief that "any lubricant will work"—that oil is oil and grease is grease, regardless of application. This couldn't be further from the truth. The reality is that choosing the right lubricant involves careful consideration of factors such as viscosity, additive packages, and environmental conditions.

Another common misconception is the idea that "more lube is better." Over-greasing bearings can lead to a host of problems, from increased energy consumption to premature failure. It's imperative to strike the right balance and follow recommended lubrication procedures.

Not all lubricants are equal. Many are incompatible and when mixed together can cause a negative change in state detrimental to the lubricated bearing and feed lines.

Lubrication has always been colored with the incorrect perception that its seeming simplicity and inexpensive nature can be successfully managed with little or no knowledge or training. This is further exacerbated when neglect and/or improper application do not result in any immediate bearing or machine failure.

Finally, we come to the undeniable truth: cleanliness is paramount. Dirty lubrication systems and contaminated lubricants spell trouble for machinery health and performance. Establishing rigorous cleaning regimens and maintaining pristine lubricant reservoirs are crucial steps in ensuring smooth operation.

As you delve deeper into the world of lubrication, do not be afraid to challenge these myths, embrace best practices, and strive for excellence in machinery maintenance. This book is designed to help you navigate the complexities of lubrication with knowledge, diligence, and a commitment to optimal performance.

As you embark on your journey through these pages, you'll find a curated collection of insights distilled from columns and articles I have written and published over the years in esteemed publications such as Maintenance Technology Magazine, Efficient Plant Magazine, and The RAM Review e-zine.

Structured into eight chapters, each chapter of this book is designed to address specific challenges encountered in the practical lubrication of assets. Whether you're just starting out or seeking to refine your existing practices, you'll discover actionable tips within each chapter, serving as springboards for immediate success in your lubrication endeavors. Each chapter focuses on specific areas known to deliver the greatest success when implemented.

Drawing from over three decades of hands-on experience working across diverse industries, the strategies and tips presented here have been rigorously tested and proven effective. It is my hope that this guide will empower you to develop and implement a best practice lubrication management program within your organization.

As you navigate through these pages, know that you are not alone in your journey. Consider this book a trusted companion, offering guidance and support as you navigate the complexities of lubrication management. I wish you great success as you embrace the challenges and opportunities that lie ahead in your lubrication journey.

Kenneth E. Bannister, 2024

Top Ten Lubrication Failure Checklist

Underestimating your lubrication approach is gambling with your machine and operational health!

This list depicts the ten major reasons for lubrication failure. The first nine reasons are directly or indirectly related to the acceleration of lubricant, component, and machine failure:

- Lack of lubrication training
- Lack of lubrication application engineering
- Poor housekeeping, i.e., lack of order and cleanliness
- Over lubrication
- Under lubrication
- Use of dirty or contaminated new lubricants
- Infrequent oil/filter changes
- Cross contamination of existing and incompatible lubricants
- Incorrect lubricant (additive package, viscosity) used on a bearing
- Bearings mounted out of square or misaligned during setup.

If you have suffered or are currently suffering lubrication related failure like these in the above list, read on to gain valuable and easy to adopt practical strategies and tips on how to "kick-start" your new lubrication program!

About the Author

Kenneth E. Bannister is a UK technical apprenticed and accredited mechanical design engineer, credited on several engineering patents, two of which involved tribology aspects in their design. Ken is also a Certified Maintenance Reliability Professional and since 1988 has consulted worldwide helping clients implement practical and meaningful asset management, reliability, and lubrication management programs. Ken is one of a few asset management consultants holding expertise and accreditation in the field of tribology, lubrication failure management and industrial lubrication application as a designated professional Machinery Lubrication Engineer.

Ken was the first consultant to assist a company through the ISO 55001 asset management certification process in North America and was part of the development team for the ICML 55® world lubrication standard. More recently Ken was a contributing author and senior editor for the compilation of the ICML 55.0, ICML 55.1, and ICML 55.2 standard documents.

Lubrication Strategies and Tips is Ken's fifth industrial lubrication book and joins his recently published Practical Lubrication for Industrial Facilities – Fourth Edition and the "Maintenance Partnerships" handbook. Ken also wrote the very successful Lubrication for Industry book. and the lubrication section for the "Machinery's Handbook" as well as other books on Energy Management and Predictive Maintenance. Over his career, Ken has published over 650 articles and white papers for numerous international maintenance magazines, with over half dedicated to the field of practical lubrication.

Ken was a founding board member of the Plant Engineering Maintenance Association of Canada (PEMAC) and currently sits on the board of directors for the International Council for Machinery Lubrication (ICML) responsible for the rollout of the ICML 55® world lubrication standard.

CHAPTER

1

Design Stage Lubrication Strategies and Tips

Chapter one addresses strategies that are best employed in the design stage of the lubrication management equipment being lubricated. It is well understood that bearing lubrication decisions, made at the machine design stage, often dictate how well the machine will be maintained throughout its working life.

1.1 Design Strategy One: Designing Lubrication Delivery Systems based on Lifecycle Costs, Maintainability, and Environmental Factors

Most design engineers understand how to correctly size a bearing for load or how to choose the proper bearing class, limits, fits, and tolerances for an application. Very few, however, are versed in the science of lubrication and basic maintenance practices. This is evident when a grease-lubricated bearing is mounted inboard of the machine side frames in a hidden or difficult-to-reach location with a simple grease nipple at the bearing point—the type of arrangement that requires a full machine shutdown and lockout/tagout to lubricate a single bearing—should it get lubricated at all.

Awareness of "in service" consequences at the lubrication program and system design stage will make the operational effort much more efficient when the machine goes into service, increase the asset life cycle, and reduce downtime incidences and their associated costs for the life of the machine.

Depending on the machine design, bearing lubrication may be performed using oil, grease, or both. Oil is prevalent with closed gearboxes, high temperature, and high-speed bearing applications. Grease systems are usually less expensive and more common. The following are the major pros and cons of a number of popular lubrication application methods:

Manual delivery: A simple method employing a $1.00 grease nipple screwed into each bearing point. The pro side is minimal cost and minimum design requirement. The cons are a lifetime of uncontrolled delivery—too much, too little, none at all; bearing access problems as they are difficult to see and access at the bearing point; and the increased chance of contamination/cross contamination from dirty grease gun tips and non-identifiable grease tubes in the grease gun.

Single-point auto lubricator: Newer versions are programmable, refillable, can be used for oil or grease, and develop enough pressure to move an eight-point divider valve, all for an investment of just a few hundred dollars. These are great for existing applications with minimum point requirements and out-of-the-way single points, but are still considered "throw away" lubricators that, in the end, cost many times more than a traditional automated-pump system.

Centralized automated single-line and dual-line injector systems: Used for oil or grease delivery; require little up-front engineering while offering built-in control flexibility at each injector point. Unfortunately, good features can also be detrimental, especially when flexible also means tamperable, i.e., easy adjustments that allow non-experts, with the best intentions, to starve or over-grease bearings.

Centralized automated progressive divider system: Probably the most respected delivery system on the market as this is a fully engineered system that is almost always included in the system cost.

Unfortunately, even with a lubrication-savvy designer who has specified a state-of-the-art fully centralized and automated lubrication-delivery system, all can be changed with the stroke of a pen. Lubrication systems, as inexpensive as they are, are much more expensive than the $1.00 grease nipples and are an easy sacrifice when spiraling project costs need to be reined in.

If you have any load on your machine bearings you will require a lubrication strategy and need to choose a lubricant-delivery method. Choosing one is much cheaper and easier at the design stage. An automated system is the smart choice because, when set up correctly, it can increase bearing life cycle by three times that of a manually lubricated bearing, while reducing your machine's energy profile.

1.1.1 Environmental factors

Bearing longevity depends on the machine operator and the maintenance crew factoring in the working environment and making appropriate adjustments. Ironically, poor working environments can be found, even in a food plant that is washed down regularly with water and caustic soda at the end of each shift. The plant may be clean but strong water spray pointed directly at a bearing can cause the grease to wash out. In such cases the bearings must be re-greased immediately or a more permanent solution implemented such as installing water shields or run-off protective guards. A similar solution is required for processes that use aggregate or sand (silica), such as in a sandcasting foundry, cement plants, or mine processing facilities.

Other working environments detrimental to lubrication include:

A multishift plant operation: Most original-equipment manufacturers base their lubrication recommendations on a single-shift operation (2000-hour annual operation). Double and triple shifts require the lubrication delivery and reservoir fill frequency to also double and triple if the line throughput remains constant.

Aggressive production schedules: Running over the design capacity will break down the lubricant much faster, requiring a ramped-up delivery schedule.

Multi-mode operation: Different lubrication strategies are required for machinery that may operate in three separate modes, e.g., full-operation mode, standby/idle mode and/or spare/redundant modes.

Temperature extremes: Wide operating-temperature ranges dramatically affect the lubricant viscosity. Very hot conditions can make the lubricant very "thin," causing it to lose its ability to stay in the bearing pressure, or load, zone, allowing metal-to-metal contact to occur. Extremely cold environments make the lubricant "thicken." Thick lubricant may stall out the pump, causing the system to dramatically slow its delivery or stop altogether until the temperature rises again. The correct choice of lubricant viscosity is essential in extremes that may require a complete lubricant changeover as the seasons change.

In every case, the onus is on the end user to understands and make adjustments that compensate for the working environment. If at all possible, the best strategy is to specify exact needs at the design stage. Machine design and working environment can significantly affect how bearings are lubricated.

TIP √ With an oil-lubrication system, ensure that the oil reservoir has easy-to-read site-level gauges installed, a desiccant-style breather, filters with go/no go change indicators depicted in, and a contamination-controlled fill system. If you already employ lubrication standard(s) in the plant, ensure that the new design complies with your standard so as to minimize spare parts and training requirements.

TIP √ Always try to install the lubrication pump and controller at or outside the working perimeter of the machine envelope. This allows the lubricator to be filled, tested, and maintained without the need to shut down and lock out the machine.

TIP √ Formerly document your current working environment by area, production process, line, and machine. Give as many environmental factors as possible that may include process raw materials, (product), ambient and operation temperatures, speeds, water, dirt, weather change, operational timetables, etc. Ensure a process is in place to update documents regularly.

1.2 Design Strategy Two: Learn how to Protect Bearings from Cradle to Grave

Poor lubrication practice has long been accepted as the major contributor in up to 70% of all lubrication related bearing surface degradation failures*, yet the actual root cause is rarely investigated and often misunderstood.

There are many elements that affect the bearing life cycle as it transitions from the drawing board to its working environment. While design, storage, installation and set-up all play important roles in a bearing's early life, effective lubrication throughout a bearing's working life has the greatest positive impact on longevity, performance, and asset uptime. Being aware of a bearing's needs in both its design and working-life stages will determine maintainability requirements and assure maximized reliability. The following synopsis follows a bearing throughout its life-cycle stages and highlights influences and elements that can affect bearing reliability. Each stage offers easy maintenance tips to mitigate chances of premature failure.

Stage 1: Manufacture

Bearings are manufactured to the highest of tolerances and are assembled and packaged in a HEPA filter-controlled environment, commonly referred to as a "white room environment". Bearings can be ordered and readied for sale with only assembly lube, pre-lubricated with standard bearing manufacturer factory grease or oil, or pre-lubricated with a customer specified lubricant of choice. Lubricated bearings are usually shipped with somewhere between a 20% and 30% cavity fill.

TIP √ To avoid initial lubricant cross-contamination and ensure the bearing is filled with the correct viscosity and lubricant type from the onset, consider purchasing non-lubricated bearings for inventory spares that can be control filled prior to use with the exact lubricant product and amount.

Stage 2: Equipment design engineering

It is the original equipment manufacturer (OEM) who designs the machine in accordance with a nominal specification or to client specifications. Bearing type, style, and size are chosen based on operating speed, design load, temperature, and working condition factors. These factors, along with financial considerations (grease nipples are much cheaper than centralized lubrication systems), determine whether the machine is to be lubricated by oil or grease. OEMs will often gladly work with a client at the specification stage to accommodate existing lubrication specification requirements.

TIP √ If possible, provide your internal engineering and/or purchasing department with an OEM maintenance-specification list of preferred automated lube-delivery systems similar in design to in-house systems (if available), and provide a list of current lubricants used in the plant.

TIP √ Request an OEM bearing schematic that locates and lists all lubricated bearings used on the equipment by type, style, bearing ID number, and bearing fit classification for maintenance and lubrication standardization purposes.

Stage 3: Inventory

When a bearing reaches its end-of-life stage it will require a maintenance replacement. To offset equipment downtime, most maintenance departments choose to pre-purchase bearings and place them in an on-site MRO (maintenance, repair, and overhaul) inventory stockroom.

Poor quality of care afforded to the bearing during its inventory stay due to handling issues (open wrapping, touching bearing surfaces, dropping on floor), exposure to moisture and dirt, exposure to floor vibration (can cause false brinelling and flat spots on bearings) can cause a bearing to significantly deteriorate prior to first-time use.

TIP √ Ensure bearings are stored in a clean, humidity-controlled environment.

Stage 4: Installation, machine setup

During installation, future bearing life relies greatly on the skill and knowledge of personnel responsible for fitting the bearing(s) and setting up the machine. Bearings forced into place using blunt-strike tools can cause immediate surface brinelling (hard surface indenting). Bearings that fit too loose or too tight due to an incorrect fit tolerance can cause surface scoring, rapid wear, and fatigue failure.

Wherever belt and chain drive systems are used, accurate motor, pulley, and gear alignment is essential if the bearing is to survive for any length of time. Angular and offset misalignment excessively and unevenly loads the bearing(s), causing wide ball-path wear on the inner race, and non-parallel wear on the outer race.

TIP √ Ensure maintainers have the proper training and tools to install bearings correctly.

TIP √ Invest in a laser-alignment tool for use on all machine driver-assembly setups.

Stage 5: Working life

Assuming a bearing has successfully arrived at its working-life stage in good operational condition, it is now ready to provide service. At this point, it is the operating conditions and the quality of the lubrication-management program that will dictate the life cycle of each and every bearing. For example, a standard 6206-RS sealed bearing employed in a pharmaceutical manufacturing plant is likely to survive longer than its counterpart in a food-processing plant or foundry. This is due to the white-room controlled environment and light duty cycle of the pharmaceutical plant in comparison to a food-processing plant, in which equipment is water washed every day, and a foundry that uses lots of heat and sand (silica) in its manufacturing process. To combat these conditions, a formal lubrication-management program must ensure that the right lubricant is placed in the right bearing, in the right amount, at the right time! If no lubrication-management program is currently in place, consider beginning your program with the following three tips.

TIP √ Introduce a lubricant-consolidation program that reviews your current lubricant catalog with the intention of updating and streamlining the number and types of lubricants used. This will reduce cross-lubricant contamination and provide better management of current lubricant stocks.

TIP √ Introduce a contamination-control program that examines maintenance cleanliness habits when handling lubricating bearings and lubricant storage, handling, and transfer practices. In more-severe process industries this program can be expanded to investigate opportunities to mitigate lubricant and bearing contamination due to process fluids, moisture, dirt, and production solids fallout.

TIP √ Provide lubrication-awareness training for all maintenance and supervisory staff and lubrication certification training supported by ICML (International Council for Machinery Lubrication), Broken Arrow, OK (icmlonline.com) and STLE (Society of Tribologists and Lubrication Engineers), Park Ridge, IL (stle.org) certification bodies for lubrication technicians and champions.

If installed correctly, operated within its design specification, and lubricated in a knowledgeable manner, a bearing will deliver a long reliable service life, often three times longer than a poorly looked after counterpart. If a bearing fails within its first six months of operation on a new piece of equipment, suspect the design or OEM installation as the root cause. Similarly, if a replacement bearing on an existing machine fails in the same time period look internally at your inventory methods and installation techniques. Bearings failing after their warranty period usually require detailed analysis of the lubricant and the physical bearing to indicate where your lubrication program requires attention.

As your lubrication-management program matures, bearing failure should drop dramatically, allowing time to assess future bearing failures through lubricant and forensic analysis of the failed bearing itself. When your next bearing fails, don't throw it away. Instead, photograph, tag, and bag the failed bearing and send it off for analysis. Simultaneously, perform an internal review to see how you compare to the analysis report.

*Dr. Ernest Rabinowitcz

1.3 Design Strategy Three: Specify and Choose the Correct Lubricant Seals for the Job

Once again, your application, operating conditions, and lubricant quality will determine which seal design will best keep contaminants from damaging rotating equipment.

If your equipment uses components that rotate, oscillate, or reciprocate, that same equipment requires lubricant to protect the moving bearing surfaces, and sealing devices to contain the lubricant ion the machine boundaries.

Sealing devices perform two functions. (1) they prevent lubricants or gases from leaking out of a reservoir or bearing surface area and (2) they minimize contaminants (water and dirt) from entering the bearing surface area. In short, seals are designed machine elements, or "gatekeeper" devices, that separate spaces containing different fluids or substances that may or may not be subject to pressure differentiation.

When asked "what is a seal," a maintainer is most likely to describe one of three types of shaft seals. Shaft seals come in several shapes, sizes, and materials and must be matched to the application, temperature, and lubricant properties.

Choosing a suitable lubricant for your application will require knowledge of the operational conditions (shaft speed, bearing load, hours of operation), the ambient working conditions (hot, cold, dirty, wet), and the type of shaft seals employed. To meet the shaft seal requirements and deliver long-term reliability, the lubricant must meet specific criteria:

- Permit damage-fee installation of the seal
- Dissipate frictional heat
- Increase sealing effect
- Prevent seal adhesion, even after a long standstill
- Permit easy assembly/disassembly
- Demonstrate compatibility with the sealing material and resistance to ambient media.

Radial-lip, labyrinth, and bearing isolators are the predominant shaft-seal designs in use today. They differ considerably in that the inexpensive radial-lip seal is referred to as a contact seal and is primarily used to isolate oil systems. The more expensive labyrinth and bearing-isolator seals are referred to as non-contact seals that can isolate lubricants and gases.

1.3.1 Labyrinth seal

In their 2009 *Journal of Mechanical Science and Technology* paper titled "Comparative analysis of the influence of labyrinth-seal configuration on leakage behavior," authors Tong Seop Kim and Kyu Sang Cha describe a labyrinth seal as "a non-contacting sealing device that consists of a series of cavities connected by small clearances [in which] flow loses its total pressure while it sequentially experiences acceleration into the clearance due to contraction, friction through the clearance, and dissipation of kinetic energy at the cavity." The cavities and small clearances create a torturous pathway that results in turbulence acting as a curtain or barrier to restrict outward flow (egress) of lubricant or gas, and inward flow (ingress) of contaminants. Because a labyrinth seal is non-contacting, it technically will not wear out and should never need replacing. See Figure 1.1, a simple labyrinth seal.

Figure 1.1: A simple labyrinth seal.

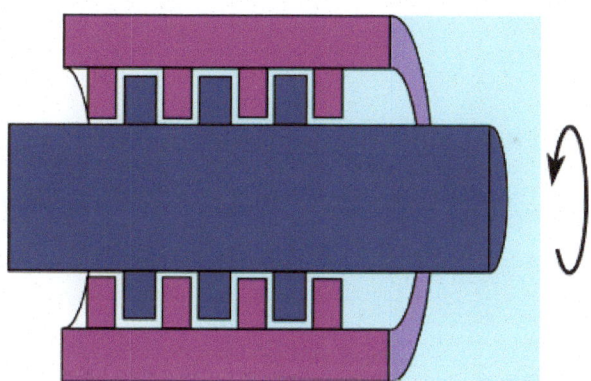

1.3.2 Radial-lip seal

A radial-lip seal is a point-contact seal that uses a circular metal outer band that captively fits into the stationary housing bore. A bonded-elastomer sealing lip, contained in the outer band, sits against the rotating shaft and provides a dynamic and static seal. Most radial-lip seals also use a garter spring to ensure that the single point contact is always engaged on the shaft. See Figure 1.2, a radial lip seal

Figure 1.2: Typical radial-lip sealed bearing in service. Note: bearing has been over lubricated as evidenced by the trace lubricant that has bypassed the seal.

Courtesy: ENGTECH Industries Inc.

When a shaft is rotating at speed, the shaft seal runs on a thin layer of lubricant between the lip and shaft. Lubricant is also hydrodynamically pumped into the areas by the centrifugal pumping action of the rotating shaft. This lubrication effect is beneficial to the life of the seal but is only as good as the oil condition. If the oil is in poor condition, the sliding contact surface will cause the seal to heat up and prematurely wear, losing its sealing qualities.

Radial-lip seals are classified and purchased according to surface-speed ranges. For example, a natural buna-N rubber elastomer is good from −40 F to 225 F and it works well with petroleum-based lubricants. A Viton seal is good from 0 F to 400 F and is excellent for all synthetic fluids that can "soften" other elastomer types. TFE (tetraflouroethylene) is a harder plastic that requires greater installation care, but is good from −100 F to 400 F. TFE is compatible with most fluids.

Although inexpensive to purchase, radial-lip seals are full-contact designs and, as such, require the seal to come into contact with the lubricant, making lubricant choice a decisive factor when choosing this type of seal.

1.3.3 Bearing isolators

The bearing isolator is a more-recent seal design that is gaining popularity. This hybrid compact design is a dynamic non-contact two-piece unit consisting of a fixed piece (stator) that interconnects and marries up to a moveable rotor piece attached to the shaft. The bearing isolator is an easy-to-install split-hybrid design that uses O-rings and an elastomer connecting ring, known as the unitizing element. See Figure 1.3, a bearing isolator cross-section.

When specifying seals, first understand your operating conditions. Then consult with your lubrication and bearing seal providers to determine which seal will best protect your assets.

Figure 1.3: Bearing isolator cross-section.

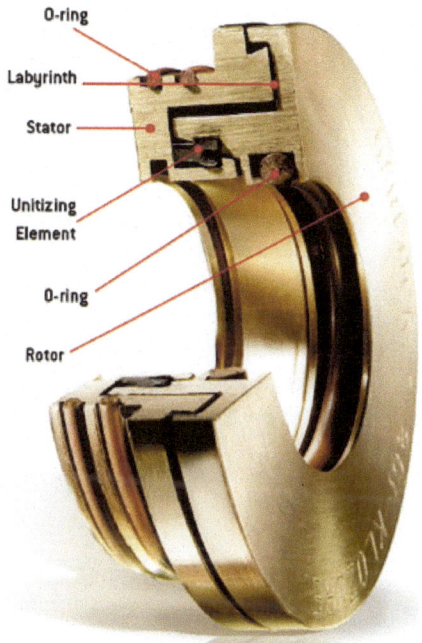

O-ring

Labyrinth

Stator

Unitizing Element

O-ring

Rotor

Courtesy: Impro-Seal

1.4 Design Strategy Four: Adopt a "Lubrication by Design" Approach Toward Lubrication

Arguably, the most used and abused instruction in the field of practical lubrication is "lubricate as necessary." The origin of this advice is most often attributed to the OEM's (original equipment manufacturer) machine operation and maintenance (O&M) manual.

Unless contracted to specifically prepare a custom O&M manual for a client's specific operating environment, the OEM will likely defer to a generic manual that uses subjective language to outline a typical maintenance approach. This is especially true when the OEM sells its equipment globally through third-party agencies and retains little control—or understanding—of how and where that equipment is used. The level of subjectivity is further amplified when an unsuspecting and/or unenlightened maintenance department unquestioningly follows the unaltered written instructions to suit their unique working environment.

A key component in reliability and performance improvement—with regard to both maintainers and machines—is consistency of effort. This type of consistency is afforded through an understanding in two major areas: (1) how a current operating environment impacts the machinery requirements and (2) who interacts with the machine to perform lubrication tasks.

Obtaining the highest level of machine reliability is best achieved by defining the simplest of maintenance observations and tasks based on the equipment's weakest links. A machine's weak links typically present themselves in two formats: consumables and adjustables. Often thought of as "nuisance" or "pain" points, weak links are instantly identifiable systems or components of an equipment system that require regular or constant replacement or modification (tweaking). Lubrication falls into both of these categories.

1.4.1 Recognizing your work environment

Lubricants are thought of as consumables because of their propensity to be consumed or deteriorate in service or leak out of a closed environment; all conditions requiring a "top-up" replenishment or full replacement strategy. A machine's working environment can dictate how quickly the lubricant will deteriorate. For example, a bearing operating in an extreme wet, damp, hot, or dirty environment similar to that found in a foundry, mining operation, or steelmaking operation, will call for a more focused lubrication approach than similar bearings operating in "white-rooms", HEPA-filtered environments such as pharmaceutical-manufacturing operations.

Lubrication-delivery systems also require monitoring to determine application requirements and schedule adjustments based on changing operational needs. For example, utilizing a manual greasing approach in bearing lubrication will require a change in PM (preventive maintenance) frequency when moving from a single-shift to double-shift operation. Similarly, in changing to an automated lubricant-delivery system, note that the lubricant reservoir will likely require replenishment at twice the previous (manual) rate and necessitate an adjustment of the lubricant-fill cycle.

Recognition of your working environment, and tailoring your lubrication approach accordingly is the first step to implementing a "lubrication by design" method and, ultimately, achieving true lubrication effectiveness.

1.4.2 Objective instruction and machine interaction

Instructing a maintainer, or operator to "lubricate as necessary" will only guarantee subjective decision making regarding which lubricant to use, how much to use, and how often it is to be applied. Subjectivity, in turn, invokes inconsistent behavior leading to cross-contamination of lubricants, over-or-under filling of reservoirs, breaching of bearing seals, over-filling or starving bearings. These situations all reflect undesirable high-risk behavior that results in premature, yet preventable, machine failure and downtime.

In Dr. Atul Gawande's 2009 best-selling book, "The Checklist Manifesto: How to Get Things Right" (Metropolitan Books, New York), he described the first military test flight, more than 75 years earlier, of the Boeing B17 bomber that had been introduced in the late 1930s. Ending in a crash due to a simple oversight by the most experienced pilot in the US Army at the time, this flight led to the aviation industry pioneering operational and maintenance checklists.

Designed to overcome human ineptitude, attitude, and ignorance, the aviation checklist, written in simple and exact language familiar to the profession, was instituted to ensure that each and every pilot from that point on followed a consistent "readiness" check procedure prior to take-off and landing. As head of the World Health Organization's "Safe Surgery Saves Lives" program, Dr. Gawande successfully adapted the aviation checklist methodology into a simple, innovative tool for the medical field—and subsequently credited its use for a dramatic reduction in hospital and surgical deaths, regardless of hospital conditions. There's a significant takeaway from this story for those of us who have an interest in the health and well-being of industrial equipment and processes.

Lubrication checklists that don't challenge or insult maintainers or operators (but are designed correctly and written in a concise manner similar to those used in the aviation and medical fields) can overcome both ignorance and ineptitude and promote low risk through a high degree of consistency.

Table 1.1: Lubrication checklist.

Task sequence	Lubrication instruction
1	Lubricate red grease points numbered 1, 3, and 7 each with 2cc (2 full grease gun shots) of ABC EP1 grease in the red-sleeved grease gun.
2	Lubricate red grease points numbered 2 and 4 each with 4cc (4 full grease gun shots) of ABC EP1 grease in the red-sleeved grease gun.
3	Lubricate blue grease points numbered 5 and 6 each with 2cc (2 full grease gun shots) of XYZ 505 grease in the blue-sleeved grease gun.
	Check:
4	Is gearbox # 12 reservoir oil level between the Hi and Lo level marker? If Yes, move on to next task. If below the Lo marker level, fill reservoir to Hi marker level with ABC ISO 220 gear oil. If above Hi marker level, not on work order and report condition to supervisor.

Take, for example, the sample checklist shown in Table 1.1. Written in objective language, it points to minor, required steps in making modifications to the lubrication-system components of a gearbox. It specifically references HI-LO-fill indicators regarding the lubricant-reservoir sight gauge (see Figure 1.4) that help personnel making a simple, yet accurate, go/no-go decisions on filling the reservoir, and a color identification of the grease nipple and grease gun (see Figure 1.5) used to visually identify the correct lubricant amount and type for each bearing.

Color, though, is just one aspect of identification that this checklist calls out. It also references the exact identification of the grease point and reservoir number, the correct grease that's to be used, and the amount of the grease to be deployed in both displacement and grease-gun shot action.

TIP √ To facilitate use of the procedures in the sample checklist (Table 1.1), implement a simple grease-gun consolidation program. The program requires the following actions:

Figure 1.4: Typical HI-LO level markings for an oil filled gearbox.

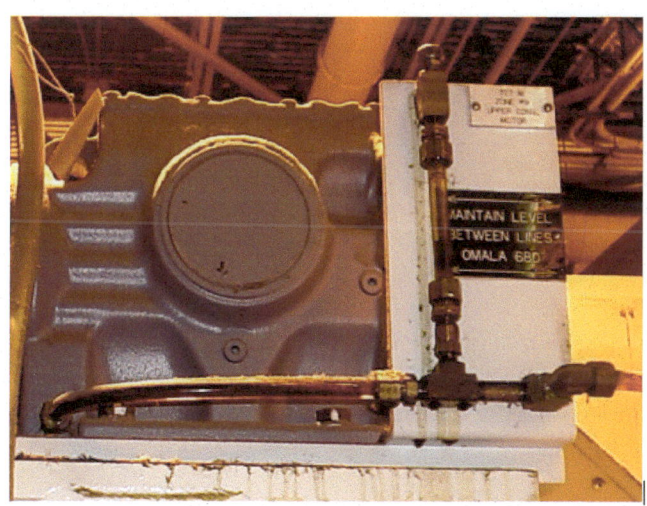

Courtesy: **ENGTECH** Industries Inc.

Figure 1.5: Color coded grease gun and grease fitting ID collar.

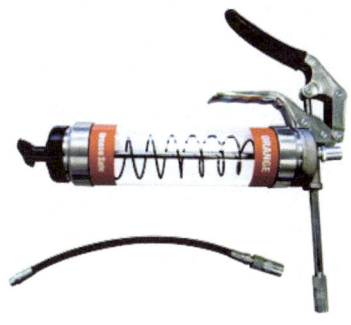 

Courtesy: Fluid Defense Systems

- Collect all current grease guns in use within the plant.
- Give these old grease guns a new home by allowing your maintainers to take them home for personal use.
- Purchase a new set of identical make, type and fixed discharge capacity grease guns.

- Establish the grease gun discharge or "single shot" capacity in volume and/or weight to recalculate the exact number of shots required for each bearing lubricated in the plant, regardless of the grease gun is used.
- If different greases are in use, employ a grease-gun and grease-point colors system.

1.4.3 Bottom line

The "lubrication by design" approach detailed here requires almost no capital outlay. With some minor organizational effort up front, it can be rolled out systematically, machine by machine, throughout the plant in a timely manner.

TIP √ According to Daniel Boorman of Seattle, WA—the person charged with developing aviation checklist manuals for all of Boeing's planes for 20+ years—the secret of a good checklist is in how it's written, starting with the use of simple and precise language familiar to a profession (users).

TIP √ A checklist doesn't have to be too comprehensive to be effective (usually between five and nine items).

TIP √ Well-designed checklists fit the flow of the specific work, encourage users to read each point out loud, and help them detect potential failures before they occur.

TIP √ A successful checklist ideally fits on one page, is free of unnecessary color and clutter and uses both upper and lower case in a sans-serif font such as **Helvetica**.

2

Lubrication Management Strategies and Tips

To successfully manage an industrial lubrication program the maintenance management team must seek out, recognize, and take advantage of lubrication work management opportunities as they arise.

Chapter 2 introduces some proven non-traditional strategies, as well as some traditional strategies that are often ignored. These strategies are inexpensive and easy to deploy in any type of industry.

2.1 Lubrication Management Strategy One: Adopting a Failure Scene Investigation (FSI) Approach to Lubrication Failure Management, Analytics, and Diagnostics

For decades, maintenance has prided itself on its ability to swoop in at a moment notice and save the day whenever a breakdown occurred. The 1970s and 1980s witnessed an industrial maintenance shift from paper-based reactive dominated maintenance models to PC computer based preventive maintenance models due to the advent of widely available, user-friendly computerized maintenance management software (CMMS) programs. With searchable code management and document storage/retrieval capability at maintenance fingertips, an improved understanding of how to prevent machine failure blossomed. Unfortunately, many maintenance improvement approaches mobilized at this time have not evolved to meet the changing needs of today's industrial maintenance environment and still dominate the present.

Although companies have updated and/or changed their CMMS programs and have added preventive maintenance (PM) work orders to the system for new equipment purchases over the years, few have actually taken advantage of the data-driven decision-making analytic capability most CMMS programs provide. It is this very capability that allows a CMMS program to state it is a management system. If this feature is unused, it is merely an electronic work order system.

2.1.1 Analytics

Marcel Proust, a 19th century novelist once stated, "The real voyage of discovery consists not in making new landscapes but in having new eyes." Analytics are the eyes that allow us to collate and turn meaningless unrelated data (MUD) into meaningful information we can make true management decisions with.

TIP √ Determine how well your CMMS can function as a management system by performing a quick query (informal report) litmus test.

Test one is a simple maintenance spend report on the current year and past two years to determine the % maintenance spend on labor versus maintenance parts. Test two: drill down further to determine the % spend on all lubrication work and materials/parts used versus the total maintenance spend. If the code management is set up well, test three can mine deeper to determine how much was spent on labor and parts through outside contractors versus internal staff and parts.

The next test is to determine over the same periods of time the mean time between failure (MTBF) for all assets combined. These figures can now be shown on a simple graph to show the gross maintenance spend relationship to overall asset reliability. These simple reports allow management to perform a validity test of maintenance spend versus reliability that can be more focused by performing similar reporting on smaller groups of assets such as manufacturing (product) line, location, equipment type/group, supervisor, etc. to discover reduced levels of performance and laser focused opportunity for improvement. This information can be scrutinized further to determine if the lubrication task or schedule is valid. The ability to perform these simple performance reports allows the user to relate the outcome to maintenance processes and convert the data into actionable information.

Unfortunately, many CMMS and enterprise asset management (EAM) systems in use today still have trouble passing the litmus test. Although they are likely charged full of data, the data is most often not relevant for reporting purposes or difficult to access due to ineffective system code management set up. Other barriers to analytics can include work that is performed but not captured on a work order or entered into the CMMS, closed work orders that do not contain vital data such as actual hours, parts used, tools used, failure codes, etc. If this is your system and you are unable to easily perform a basic litmus test your CMMS is likely wallowing in MUD. Ironically, MUDA is the Japanese word used for waste when implementing a lean approach to production and maintenance.

2.1.2 Diagnostics

The standardization of crime scene photography has changed little since its introduction in 1888 by French police officer Alphonse Bertillon.

Many bearings continue to suffer an early demise due to poor implementation, maintenance, lubrication and operator practices. Only when we understand the "how" and "why" an asset fails in storage and operation are we able to truly prevent their premature failure. This can only be understood by allowing a standardized failure scene investigation (FSI) process designed to take place immediately following a failure event. The FSI always commences with a photo assay of the "failure/crime" scene in an identical manner to the process Alphonse Bertillon followed all those years ago.

Today's communication devices and cameras offer no excuse to not photograph a failure scene captured against an investigative work order file in the CMMS. Each time a piece of equipment or component fails it leaves behind an evidence trail that will lead not only to the failure cause, but also deliver a strategy to understand and predict/prevent future failure event occurrences.

If we are to reduce our levels of maintainability while increasing both availability and reliability it behooves maintainers to develop a systematic approach to diagnosing a failure scene that follows the FSI lead by commencing with photographing and documenting all contextual aspects of the failure scene and not destroy the failure scene by contaminating or throwing out evidence in our haste to "save the day". These investigation documents and components are essential for forensic and failure analysis and planning and scheduling use.

TIP √ Start performing failure scene investigations immediately using the following simple seven-step method:

1. Secure the scene—work with machine operator to perform a quality evaluation of the failure scene before commencing repairs and/or restarting the equipment.
2. Photograph the scene—the old adage "a picture is worth 1000 words" could not be truer in a failure investigation. Photos allow the failure scene to be revisited well after the equipment is back up and running, and act as good training materials for preventing future failures.
3. Perform on-scene diagnostics as required—the maintenance reliability group can perform many technical diagnostics at the failure scene that can include infra-red signatures, oil analysis signatures, metallurgy, etc.
4. Bag and tag all physical evidence of failure or tampering—once all local physical evidence of tampering and breakage has been photographed, tagged and bagged, the actual failed components can be dismantled and replaced. Any parts for repair must be photographed, and any parts requiring replacement must also be bagged and tagged.
5. Interview witnesses—operators can describe any abnormal sound, smell or vibration emanating from the equipment prior to failure.
6. Perform laboratory testing—examine all past failure records and vibration readings, performing any necessary metallurgical and oil testing.
7. Analyze findings, writing up a root cause analysis of failure (RCAF) report c/w recommendations and update preventive strategy as required.

Adopting an FSI approach to failure diagnostic investigation is sure to enhance your lubrication and reliability program and improve your asset management approach.

2.2 Lubrication Management Strategy Two: Take Advantage of Planned and Unplanned Shutdowns

Shutdown and turnaround events are often confused and, depending on the type of industry, can have the same or different meanings. Both are, however, meticulously planned, short-duration, (days to weeks) events that can cause large-scale planned cessation of work within a plant.

Historically, maintenance has defined shutdown events as unplanned equipment-failure situations that cause an operational production line, process, area, or section of a plant to be temporarily turned off or closed for emergency repair. Normal operation usually resumes immediately following the repair of the failed equipment.

Turnarounds are different in that they are traditionally defined as planned downtime events that require the closure of an entire operational plant, or facility, to perform one or many pre-planned technology or system upgrades, equipment upgrades, and maintenance restorations, within a defined time period.

A third event, typically known as an outage, can be either a planned or unplanned event that can close all or a portion of a plant, i.e., process, line, machine, area, or section. Unplanned outages are typically caused by raw-material, manpower shortages or power failures. Planned outages are more common in lean-manufacturing environments when production totals have been achieved for the day or for the run.

From a lubrication-program-management perspective, shutdowns, turn-arounds, and outages all present excellent opportunity windows to perform a variety of jobs that ordinarily would be difficult to perform on assets that are operating in production mode.

TIP √ Perform a cleanliness blitz. Quite simply, lubrication systems and the machines they serve are not dirt tolerant. Lubricant-wetted surfaces readily attract dirt that can easily make its way into the bearing surface areas and significantly reduce their life expectancy. Clean machine surfaces, lubrication-system reservoirs, and delivery-system components facilitate the troubleshooting and elimination of leaks. Damaged lines are easily replaced, loose fittings can be tightened, and difficult-to-reach inboard bearings are more accessible.

Perform an oil change. Now that the machine and lubrication system have been given a clean, fresh appearance, why not perform a simple oil and filter change? As basic as this sounds, this simplest-of-all preventive-maintenance strategy is often overlooked during machine downtime.

TIP √ Consolidate your lubricant inventory. Prior to a planned turnaround or outage, consider performing a lubricant-consolidation analysis. Also, use the machine downtime to flush and change out old machine lubricants in favor of the newer consolidated lubricant choices. Reservoir and pump lubricant labels should be updated at the same time.

TIP √ Modernize, standardize, and automate your lubrication-delivery systems. Perform a lubrication-system upgrade can substantially reduce maintenance effort, increase machine reliability, and lower energy consumption. Upgrades don't have to be all encompassing the first time around. Consider an incremental upgrade program based on machine availability and cost. For example, the following upgrade suggestions demonstrate how a simple grease-point system can incrementally graduate to a full-blown automated delivery system in four steps:

Step 1: Upgrade individual grease nipples at bearing points to a single grease nipple attached to an engineered progressive divider block piped to all grease points.

Step 2: Upgrade the progressive divider block system by removing the single grease nipple in the block and attaching it to a fixed manual crank-arm grease pump and reservoir assembly.

Step 3: Upgrade/replace a manual crank-arm grease pump with an automated grease-pump and reservoir assembly operated autonomously by a programmable electronic controller.

Step 4: Upgrade an automated grease system by integrating system control and alarms into the production SCADA (supervisory control and data acquisition) system or machine-control system set to trigger events and work-order requests in the lubrication work-management system software.

Oil-lubrication systems can be upgraded in a similar manner, with the following additions specific to oil:

- Upgrade recirculating-oil systems to include engineered oil-sampling ports that allow sampling to be performed live in a consistent manner.
- Upgrade reservoir sight glasses to include HI-LO fluid-level markers to ensure the reservoir fill level is in the desired fill range.

2.2.1 Be prepared

The key to taking advantage of any opportunity is to be prepared. Turnarounds are the larger planned events that lend themselves well to major upgrades of the lubrication systems. These require more up-front system engineering, planning, and parts purchase prior to execution. Major upgrades may require that parts be staged in preparation for release.

Due to their unplanned nature, shutdowns and outages tend to require an immediate action response, as their duration is often short lived. To take advantage of the situation requires flexibility and ability to tailor a job plan to match the estimated available-time window to perform work.

Regular reviews of current lubrication related issues lead to improved analysis and understanding of the required system/program needs. Of course, any changes to a lubrication system will affect its current PM work order(s), requiring assessment and updating to accommodate the planned upgraded system design.

If improvement/upgrade work is to be performed during a planned outage or turnaround, there will be a need for the following:

- Secure, accessible, temporary lay down or staging area(s) for all materials and parts
- Additional parking requirement for additional persons on site
- Permit requirements for items such as hot work or confined space
- Insurance certificates for all contracted staff
- Shutdown and startup procedures for all affected equipment.

Lubrication-system improvements make a lot of sense. If improvement can be designed and implemented with little or no impact on the operations, they make even more sense. Shutdowns, turnarounds, and outages present non-traditional opportunities that, in the past, have often been overlooked. The turnaround manager or production planner must be engaged to determine when the event is to take place as multiple projects can be performed concurrently and consecutively during the turnaround availability window. The turnaround manager and the turnaround team will determine the suitability and priority of each project to build an effective turnaround project plan. To improve chances of having your lubrication project included in the turnaround plan consider providing the following:

- Number of internal or contracted staff working on or around a machine during the shutdown process.
- Work orders with detailed, objectively written, work instructions for all work to be performed on the lubrication upgrade.

TIP √ Implement a simple pit-stop lubrication program. Have on hand a series of "schedule ready oil change or grease job" plans that can be executed immediately should a stoppage opportunity occur.

TIP √ When a shutdown, turnaround, or outage occurs, if possible, use the opportunity to perform a cleanliness blitz to facilitate troubleshooting and to find and eliminate leaks.

TIP √ Commence improvement of any grease nipple lubricated assets by bringing all of the grease points into an engineered progressive divider block piped to all grease points. This can then be automated to an auto lube pump system once funds allow.

2.3 Lubrication Management Strategy Three: Prepare for a Certification Audit

Any opportunity to review and update your lubrication practices should always be embraced on behalf of the well-being of your equipment and the environment! A well-managed lubrication management program is an integral part of any asset management program that can seek recognition through a number of relevant world class industry standards that include the following (Figure 2.1):

- ICML 55.1® Lubrication management standard (International Council for Machinery Lubrication)
- ISO 55001 – Asset management standard
- ISO 50001 – Energy management standard
- ISO 14001 – Environment management standard
- ISO 45001 – Occupational health and safety standard
- ISO 22001 – Food safety standard
- ISO 9001 – Quality management standard.

Figure 2.1: The ICML 55.1® Lubrication Standard aligns and supplements many other ISO standards.

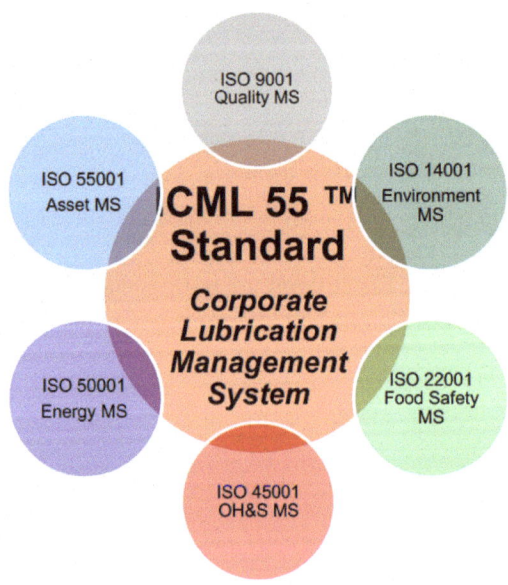

Courtesy: ENGTECH Industries Inc.

Preparing for certification to a standard will require an internal audit to review the validity and value of the current lubrication/maintenance strategies, methods, processes, procedures, etc. to ensure they align with the ICML and ISO standard requirements. ISO 55001 demands that the asset management group's approach to maintenance directly align with corporate values, goals and objectives. For example, the company may have stated holistic values and objectives that include energy use/savings and environmental sustainability edicts. The company will also have business objectives that could include increased service levels, increased manufacturing throughput and reduce capital spending and/or operating costs.

To certify, the maintenance department must clearly demonstrate corporate alignment of its asset management approach (known as the asset management system) that includes maintenance policy, it's strategic asset management plan, asset management goals and objectives, and the development and implementation of maintenance plans and reports used to validate the system effectiveness. ISO 55001 checks to ensure the maintenance department takes a value-based approach toward its assets to assure asset dependability (availability, reliability, maintainability, and maintenance support) and life cycle costing/management measures.

In a similar manner, The ICML 55.1® standard ensures lubrication practices are a major factor in increasing an asset's measure of dependability and increasing life cycle of both the physical asset and its moving parts. The hallmarks of a best practice lubrication program are those designed to meet the needs of the asset(s), to improve the maintainability process, to increase production/operations quality and throughput, and assist in the reduction of corporate energy use and carbon footprint with minimal capital outlay. Standards provide a great opportunity to review your current lubrication practices and implement an integrated and effective lubrication management program designed to help you meet your certification requirements and more importantly, help you better serve your corporation, client and asset needs!

TIP √ Engage a lubrication management consultant/expert to perform a basic lubrication effectiveness review to measure your current state of lubrication against the ICML 55.1® standard requirements and develop your continuous improvement strategy and program based on the gap analysis findings.

2.4 Management Strategy Four: Using Key Performance Indicator (KPI) Measures to Showcase Lubrication Program Effectiveness

When mechanical/hydraulic devices/systems fail, the majority of their failure modes can be directly, or indirectly, attributed to ineffective lubrication practices.

In practice, direct failure modes are most often attributed to the application of too little lubricant; too much lubricant; or incorrect lubricant applied to bearing surface areas. Indirectly, failure can be attributed to lubricant cross-contamination (mixing of two or more incompatible lubricants), and poor control of solids (dirt) and fluid (water) contamination, which, when allowed to infiltrate lubricants and bearing surface areas, is responsible for rapid surface wear; seal deterioration; and major detrimental change to a lubricant state and its efficacy. Remaining failures are primarily due to design issues (overloading), assembly, or set up issues (warranty related failures). Assessing the validity and success of any lubrication program therefore, requires the capture and analysis of performance data surrounding the maintenance and availability of lubricated systems and assets.

Peter Drucker, the legendary 20th century father and inventor of modern (business) management, wrote "If you want it, measure it. If you can't measure it, forget it." Drucker was an untiring advocate of workers knowing who they are, what they do, and the impact— positive and negative—that their direct and indirect efforts regularly impose on organizations, colleagues, stakeholders, and the environment.

2.4.1 Benefits of performance measurement

A maintenance department represents one of the single largest controllable costs within any industrial operation. Poorly managed asset maintenance— in which lubrication plays a major role—will result in reduced throughput and equipment life that translate into waste and unnecessary cost. That's why accurate measurement and trending of maintenance performance is so crucial. Performance measurement generates many immediate benefits for a maintenance department and, by extension, the entire business, that include:

1. The ability to orchestrate and align corporate and departmental strategic direction through improvement programs that include a lubrication-management program.

2. Improved utilization of a CMMS (computerized maintenance management software system), or an LMS (lubrication management software) system set up to deliver meaningful reports through the use of KPIs.

3. Fast recognition of improvement opportunities as well as fast recognition of existing excellence that can be exploited.

4. Establishment of a diagnostic baseline measurement from which to set target goals, develop improvement strategy, and trend continuous improvement.

5. The ability to benchmark with other businesses.

The ISO 55001 Asset Management Standard and ICML 55.1® standards bring Drucker's philosophy to the forefront as they emulate, incorporate, and support many of his ideas within their framework requirements.

Most maintenance departments are instinctively aware of their impact on the organization and understand that a proactive approach leads to equipment reliability and availability, which leads to reduced downtime and increased or sustainable production throughput. Unfortunately, instincts often aren't always good enough.

Countless maintenance organizations have trouble measuring and accurately defining their impact on themselves and the rest of the business—despite having easy access to data through their work-order-management systems. Instead, they continuously collect meaningless unrelated data (MUD) in the mistaken belief that the data might eventually be of value. In reality, data will only become information (and useful) in two ways: when mined and translated with context and meaning, and when used to make management decisions.

In today's data-rich environment, performance-indicator reporting, typically based on KPIs (key performance indicators) is used to translate and make sense of data. KPI reporting delivers tangible evidence of change and allows users to assess the value and validity of their program(s) and the work that they, and others who work with them, perform within a time-based framework.

Performance measurement plays a major role in assessing the success of any management program. For example, Recently, the ISO 55001: 2014 Asset Management standard devotes its entire section 9.0, to performance evaluation.

Section 9.1 Monitoring, measurement, analysis and evaluation, requires the organization (in this case the maintenance department) to determine what to monitor and measure; methods for monitoring, measuring and results analysis (to ensure validity); and when to perform measurement, reporting, analysis and evaluation so as to meet stakeholder expectations in the form of asset availability, reliability and minimized cost. Figure 2.2, taken from the ISO 55000:2014(E) standard, illustrates the importance of key performance

Figure 2.2: Importance of key performance indicators.

9.1.1.3 *A set of performance indicators should be developed to measure the asset management activity and its outcomes. Measurements can be quantitative or qualitative, financial and non-financial. Indicators should provide useful information to determine both successes and areas requiring corrective action or improvement. The organization should consider the relationship and alignment between performance indicators.*

9.1.1.4 *The asset management system should employ data from monitoring and measurement to identify patterns and obtain information regarding its performance. These data should be used to evaluate whether the organization's policy and objectives are being achieved, as well as identifying corrective actions and areas for improvement.*

Source: ISO 55002:2014(E), Sections 9.1.1.3/4

indicators and how they are used to evaluate maintenance/lubrication program success, alongside the aligned achievement of departmental, corporate and stakeholder objectives.

2.4.2 Key performance indicators (KPIs)

As illustrated in Figure 2.2, KPIs allow us to set an end goal target (objective), based on multiple business needs (objectives), and track if we are successful in meeting or surpassing them. Furthermore, KPIs allow us to analyze and understand, in laser-like fashion, where we must focus our improvement efforts.

Managing with performance measurement means that reliable data must be collected on a regular basis. The obvious maintenance conduit for tracking performance is the work management system in which assets, type of work performed, and failure types are tracked simultaneously and reported on. Lubrication work management can be tracked in a dedicated lubrication management system (LMS), or can be tracked as a work type within a maintenance management system such as a computerized maintenance management software system (CMMS), enterprise asset management (EAM) system, or an information management system (IMS). KPIs rely on system reports pertaining to labor, work type, occurrence events and cycle, time, cost, inventory, trend alarms, etc. to populate the KPI data requirement. In addition to program performance, KPIs also test how well a software system is implemented and utilized by the quality and accuracy of data extracted.

Once the work management system is able to produce the data required by the KPI calculation(s), reporting routines can be set up in the system to facilitate report updates on a regular basis.

Where no dedicated lubrication management software system is employed to singularly track lubrication related work, identifying lubrication related work is simply a matter of capturing and classifying all lubrication work orders as a specific lubrication work order type. For example, many maintenance systems typically sort their work orders, prior to release, into lubrication (work specific to lubrication system maintenance and servicing, filling of reservoirs, filter changes, etc.), mechanical, electrical, facility, and general/service maintenance related work.

Failure occurrences are tracked based on the same classifications, and in addition, are tracked on work completion to determine if failures are maintenance caused/related issues (issues that maintenance/lubrication technicians exercise full control over), or non-maintenance caused/related issues (issues that maintenance and the lubrication technician cannot prevent but must manage as part of their mandate. For example: graffiti, operator damage, denied access to asset, etc.).

TIP √ Track failure consequences by coding and tracking results of any breakdown or corrective work performed into:

- No loss of service
- Minor service loss (e.g., <one shift)
- Major service loss, (e.g., >one shift).

To understand and report on the effectiveness of the lubrication program the cost of all production downtime loss and the reactive cost of all corrective, repair and breakdown work related to bearing and bearing surface components (seals, pistons, valves, ways, etc.) failure/loss must be compared to the cost and effort of the proactive work related to the filling, cleaning and maintenance of lubrication delivery systems, hydraulic fluid systems and management of lubricants. Almost all maintenance related failures are preventable!

KPIs allow a department or corporation to turn simple data collected on a work order into meaningful management information. This is often referred to as known as "data-driven decision making". KPIs are primarily used to assess risk, performance and cost; they are further categorized into, strategic KPIs (long term), tactical KPIs (mid-term) and operational KPIs (immediate).

2.4.3 Calculating KPIs

The following KPI calculations represent commonly used lubrication KPIs. KPIs can be expressed in terms of a single equipment piece or rolled up hierarchically to represent the line, area, product, plant, etc.

KPI 1 – Asset(s) (machine/line) availability (based on lubrication related failures)

[Total # lubrication hours from date A to B – (total minor + major service loss hours from same date A to B) × 100]/(Total # lubrication hours from date A to B)

KPI 2 – Lubrication related maintenance cost as a % of total maintenance cost

Note: Can be calculated for just Lubrication PM work or all Lubrication related work

[$ cost of all lubrication related maintenance work (parts and labor) from date A to B × 100]/(Total $ cost of all maintenance work (parts and labor) from date A to B)

KPI 3 – Lubrication related maintenance cost as a % of total operating cost

[$ cost of all lubrication related maintenance work (parts and labor) from date A to B ×100]/(Total $ department or corporate operating cost from date A to B)

KPI 4 – Mean time between lubrication failure (MTBLF)

[Total operational time (shift/day/hours/minutes) from date A to B]/(Total # of lubrication related failures from date A to B)

KPI 5 – Lubrication PM (LPM) work order compliance (can also be used with other lubrication WO types)

(Total # issued LPM WOs from fate A to B – total # incomplete LPM WOs > X days scheduled)/(Total # issued LPM WOs from date A to B)

KPI 6 – Lubrication work order aging report

Total # of lubrication work orders open for XX days (30, 60, 90, etc.) since issue date

KPI 7 – Lubrication work spend analysis report

Total $ cost of all lubrication related work from date A to B

KPI 8 – OLE – overall lubrication (program) effectiveness – (production based)

OLE is a derivative of OEE (overall equipment effectiveness). The only change in the calculation is in respect to availability based on lubrication related failure as opposed to all maintenance failures.

Lubrication related availability $\times$ Rate of throughput $\times$ Rate of quality

OLE can also be expressed in terms of a % of OEE to demonstrate the overall effectiveness of the lubrication program on the manufacturing process.

The old adage "What gets measured, gets done" still holds true today! Measuring your lubrication program performance allows everyone to intimately understand their business, and the impact they have on the business day to day.

TIP $\sqrt{}$ KPIs can be positioned a in such a manner that they provide vital performance information in different management perspectives using the same data. These can include:

Financial performance. These KPIs analyze the relational costs of services and functions. Typical financial indicators can include lubrication cost expressed as a percentage of total maintenance or operating cost.

Efficiency/effectiveness performance. These KPIs analyze program and department effectiveness, and areas affecting cost expenditure. MTBLF (mean time between lubrication failure) is a typical efficiency/effectiveness KPI.

Stand-alone performance. These KPIs must be broken down further or compared with other performance indicators to understand their true meaning. Consider, for example, the lubrication PM compliance indicator. By itself, PM (preventive maintenance) compliance is a weak KPI in that it doesn't analyze effectiveness of a PM program or maintenance department. Considering this type of indicator in conjunction with MTBLF, however, leads to a better understanding of compliance relevance.

Correlated performance. These KPIs allow users to view specific information in different ways. For example, in an organization that wants to focus on program direction for all departments regarding equipment availability, it would be important to understand that the inverse of a KPI of 80% machine availability is 20% machine unavailability, or downtime, if the selected machine is required for production.

2.5 Lubrication Management Strategy Five: Introducing New Lubricants into a Plant

New lubricants are introduced and sold into the plant environment every day with seemingly no consequence. The decision to purchase and try a new lubricant is usually influenced by a purchase cost reduction or purchase bid program; a new equipment purchase whose manufacturer specifies a specific lubricant not currently stocked on site; a specialty lubricant brought in by the engineering group (or sold) to try and solve an equipment problem; or through some lubrication management initiative.

In most cases, the majority of new lubricants are introduced in an informal non-controlled manner with little or no communication between the maintenance/reliability department, engineering department and/or purchasing department. Often, little or no thought is given regarding the impact the new lubricant can, and will have, on the storage, handling, maintenance and manufacturing operations of the physical plant.

With no structured lubrication program in place, the mixing of lubricants, both greases and oils, can be prevalent, and is a major contributor to premature lubricant and bearing failure due to the cross contamination of base oils and/or additive packages. For example, a lubricant containing acidic additives added to a lubricant containing base or alkaline additives can very quickly "neutralize" a lubricant's effectiveness and protection ability. This simple oversight will often result in catastrophic failure events. Anyone who has toiled over the implementation of a formal lubrication management program knows that allowing new lubricants into the plant environment must be formalized and controlled.

An essential part of any quality lubrication management program is the initial lubricant consolidation process that reviews and documents all current lubricants on site, where they are used, and how they are stored, handled, transferred and delivered to ensure lubricant and subsequent bearing contamination is minimized. This essential engineering process, performed by the lubricant manufacturer, looks for opportunities where more modern, often less expensive lubricants can be standardized for use across the plant to replace all redundant, unsafe, and out of date lubricants and minimize the number of lubricants required to operate the plant safely and effectively. In many plants, the number of lubricants stocked and used after consolidation can be less than half the original count! In doing so, the consolidation process must determine all possible lubricant compatibility issues and propose suitable engineered lubricant change out/flushing operating procedures for the standardization process to commence. Once the list of new lubricants is finalized, the following procedure must be carried out to formalize the program:

1. Prepare a formal approved lubricant list for purchasing and set up a blanket purchase order for approved lubricants.
2. Inform all affected stakeholders of the impending change to an approved lubricant list.
3. Remove all non-approved lubricant stock from the plant.
4. Develop a stock rotation/control procedure for all approved lubricants.
5. Attain up to date lubricant safety data sheets (SDS) for all approved lubricants and remove all non-approved SDS sheets.
6. Purchase dedicated (color coded) storage and transfer equipment for all approved lubricants.
7. Purchase labels for all approved lubricant reservoirs.
8. Change all lubrication filters.
9. Develop lubricant change out flushing procedure and systematically change out all non-approved lubricants in all machine reservoirs—re-label reservoir.
10. Update lubricant inventory control software: lube specification, supplier, manufacturer, code numbers, min/max levels, inventory turn rate, etc.
11. Update affected PM job tasks in CMMS to reflect new lubricant change.
12. Update any recommended changes to PM schedule in CMMS.
13. Update equipment manuals to reflect new lubricant change.
14. Update bill of materials in CMMS.
15. Update changes to lubricant disposal procedure.
16. Update any changes to reporting requirements in CMMS.
17. Perform staff training for change awareness, product handling and safety issues, product disposal, etc.
18. Inform production.
19. Develop a new lubricant trial/approval procedure for any non-approved lubricant introduced into the plant.

Once the consolidation program is implemented only approved lubricants can be brought into the plant for regular use. This does not however, exclude a new lubricant from introduction into the plant on a trial basis. Should a new lubricant trial be required, a formal request must be made to the maintenance/reliability group through the completion of "lubricant trial request form" who in turn will oversee the lubricant trial.

2.5.1 Typical trial request form attributes

A good trial request form should have enough relevant information to enable the trial to take place and collect enough relevant data from which a yes/no approval decision can be made upon trial completion; the form must answer all of the W5 – who, what, when, where, why and how questions, then document the results of the test dividing into the following seven sections of the form.

The *who* section of the from contains the name, title, department and contact details of the trial requestor; details of the lubricant supplier and manufacturer name and primary contact person details as well as details of the person, title and department performing the trial.

The *what* section of the from contains the trial lubricant specification data that will include its name, oil or grease, base oil type, viscosity, VI rating, additives, virgin oil sample datasheet #/attachment, MSDS sheet, expected compatibility issues with other approved products, seals, production raw materials, etc.

The *when* section of the from contains the expected trial duration, commencement and completion dates.

The *where* section of the from contains information of the equipment type or specific equipment number of the machine on which the lubricant is to be tested.

The *why* section of the from details why the lubricant is on trial, in what way it will benefit the trial equipment and details of the expected results such as temperature reduction, energy reduction, life increase expectation of lubricant and/or bearing surfaces, sustainability, etc. bearing failure reduction what the trial is expected to accomplish.

The *how* section of the from documents the actual test procedure specifics, including lubricant disposal after the test and the conditions to be tested for. For example: amperage draw, temperature of bearings/lubricant, lubrication system pressure cold and hot running.

The *results* section of the from details findings data and conclusions relevant to the test, including before and after data readings, photos, infrared images, vibration readings, risk vs. benefit analysis, the return on investment statement, and a recommendation for approving or not approving the lubricant for purchase and use in the plant.

Before proceeding with any lubricant "trial" ALWAYS consult with the approved lubricant manufacturer(s) to establish if they have already compatibility tested the trial lubricant with your approved lubricants and if they as the approved lubricant supplier have a comparable product you already have in stock or available to test. You should also contact the trial lubricant manufacturer and ask if they have any compatibility tests with your approved lubricants. If no testing has taken place, you can inquire if either party is willing to test compatibility on your behalf.

TIP √ When performing a lubricant trial communicate this to the plant by placing a placard or sign on the machine stating "machine under test with new lubricant" naming the lubricant under test. Make the operator aware of the test and to notify maintenance of anything unusual regarding noise, vibration, smell, leakage, etc. during the test procedure.

When no compatibility information is available or forthcoming, and you are unable to establish compatibility, you can perform your own compatibility testing as follows:

- Take samples of both oils and blend three mix samples in a 50:50, 90:10, and 10:90 ratio.
- Send to an oil analysis laboratory and have them test for filterability, sediment, and color/clarity. In addition, have the lab perform a RPVOT (rotating pressure vessel oxidization) test to determine an oil's resistance to oxidation and a storage stability comparison.
- For accurate results the test should be performed three times and the results normalized.
- Ask the lab to assist you in determining any cross-contamination risk.
- Share the results with the new lubricant manufacturer and ask for a change out/flush procedure.

Because an RPVOT test can be quite expensive to perform, in the case of non-critical equipment you may just choose to have the new lubricant manufacturer recommend a neutral flushing oil and forego testing if there are not that many lubricant changeovers to complete.

When dealing with greases, a similar process is followed in which samples are mixed in a 75:25, and 25:75 ratio and sent to the lab to test for consistency, dropping point, and shear stability.

If the trial is successful, and none of the existing approved lubricants can perform the job, the new lubricant can then be accepted as an "approved" lubricant requiring the maintenance/reliability group to once again go through the appropriate 19 steps listed above to formally integrate the new lubricant into the plant.

TIP √ Consider putting in place a toxicity program that ties in all lubricants and chemicals (maintenance and operations) brought in to the facility.

TIP √ Build an annual work order in the PM system to update the lubricant list and SDS information.

TIP √ File your approved lubricant and chemical list with your local fire and emergency response departments so they are able to put together an emergency response plan (ERP) in the case of a fire or accident at your facility.

2.6 Lubrication Management Strategy Six: Lubrication Mapping

The standard 4R principles of effective lubrication dictate that we place the *R*ight lubricant, in the *R*ight place, in the *R*ight amount, at the *R*ight time. Performing a plant wide lubrication mapping exercise allows us to determine the first three Rs, and influence the fourth R.

2.6.1 What is lubrication mapping?

Lubrication mapping is long recognized as a basic fundamental building block essential to any best practice lubrication management program, although not new—LM was first introduced by the automotive companies in the early 1900s— the practice is no longer common in a modern-day plant.

Classic lubrication mapping (LM) is the practice of identifying, classifying, and charting out the physical location of all filter, reservoir and lubrication delivery points found on each machine (industrial), or vehicle drive train/chassis (automotive) used to facilitate an end user lubrication regime as shown in Figure 2.3.

The present-day interpretation of industrial LM has been expanded to include lubrication resources (lubricant storage, transfer, delivery, filtration, environmental spill and safety item product inventory), lubrication workflow routes, and line identification marking for all centralized machine delivery systems.

Figure 2.3: 1930s Ford Model "A" lubrication map/chart.

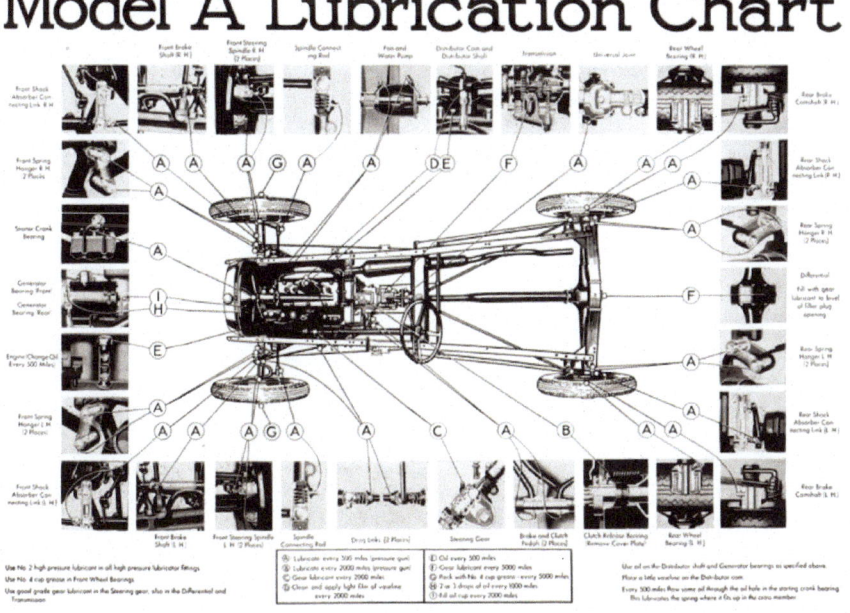

2.6.2 Benefits of lubrication mapping

Where asset reliability is concerned, it is no longer acceptable to state, "lubricate as necessary" on any machine PM work order. Building and executing an effective lubrication PM job task requires the job planner and the lubrication technician to understand each machine's lubrication requirements, alongside the logistics of efficiently managing and delivering lubricant to the required bearing point to satisfy the 4R principles of lubrication. LM is an inexpensive and highly cost-effective initiative that facilitates the commencement of a lubrication management program and delivers many benefits that:

- Provides a documented "as built" inventory and location schematic of each machine's lubrication points (Right place) and requirements (Right amount) used to build the PM job task and determine the appropriate lubricant delivery system choice if no centralized or automated system is already in place.
- Facilitates the consolidation of lubricants (Right lubricant) that results in:

- o The correct lubricant choice for the end user operating conditions
- o Reduced lubricant inventory savings
- o Reduced lubricant storage real estate requirements
- o Reduced lubricant purchasing and handling costs
- o Reduces incidence of lubricant cross contamination.

- Facilitates training and rapid assimilation of new hires to perform lubrication in a consistent and correct manner.
- Facilitates the development of efficient lubrication routing (the "ganging" of multiple lubrication PM work orders based on use of similar lubricant type to increase lubrication technician "wrench time" effectiveness).

2.6.3 Implementing a lubrication mapping initiative

If your company or department has performed a lubricant consolidation exercise in the recent past, you will likely already have much of the information required to complete a full LM initiative. Prior to any mapping exercise gather up as much information about the lubrication systems and points on the current machine inventory, these will include:

- Lubrication schematic drawings or take offs (operation and maintenance, O&M, manual)
- Machine engineering drawings (engineering department)
- Machine bill of material (BOM) lists (asset management work order system)
- Vibration monitoring route plans (maintenance planning department—these identify critical bearing point locations)
- Oil sampling route plan (maintenance planning department)
- Machine inventory list (maintenance planning department)
- Lubricant (oil and grease) purchase list (stores or purchasing department)
- Lubricant filters, breathers, spill kits, and safety items used on lubrication PM work orders (stores or purchasing department and maintenance planning department)
- SDS sheets for all lubricants.

The next step is to develop a data gathering form to document and collect the data for each machine. You will need a digital camera and a machine marker to complete the data collection. Prior to inspection, ensure the machine is de-energized and locked out and wear appropriate safety gear. Collect the following machine lubrication data:

1. Machine data such as asset number, description, and location. Photograph the machine from all four corners, four sides, underneath (if a maintenance pit exists) and above, if possible.

2. List all reservoirs for gearboxes, auto-lube systems, or hydraulics. Note if there are tags identifying lubrication information. Photograph each reservoir and note if there is a min-max fill level indication; identify any installed breathers and filters; identify fill port and drain port locations; identify reservoir capacity (measure and calculate if not known), and identify any oil analysis points.

3. Identify, photograph, and with a machine marker, number all bearing points on the machine. Note the bearing type and size and if possible, the running speed. Identify if bearings are oiled or greased and what lubricant is currently in use; if on the PM work order, list the current amount of lubricant asked for.

4. Identify if bearings are lubricated automatically, semi-automatically (centralized with manual pump or grease gun) or manually. Furthermore, check if lube lines are marked or tagged for identification and troubleshooting purposes.

With machine mapping completed, stage two can commence in which we inventory and document all lubricants found inside and outside the plant, whether in current use or not, and photograph their labels. Concurrently we can also identify, document and photograph all stocked breather and lube filters.

With stage one and two complete, all the hard work is done and the data can be turned over to your lubricant supplier who can engage the lubrication manufacturer's engineers (usually at no charge if you purchase their lubricants) to perform a consolidation audit to reduce your lubricant inventory. Once completed, this data can then be offered to one or more lubrication system suppliers, who can then analyze your machine and consolidated lubricant data to give you options and costs to optimize our lubrication delivery system designs for machine efficiency and cost and lifecycle effectiveness. A quality supplier will provide system engineering and system schematics with numbering protocols as part of their service. Similarly, approach your filter and breather suppliers in the same manner.

To develop and update stage three processes, routes and workflows, you may opt to tackle it yourself or at any time through the LM process you might want to employ a lubrication management expert who will be able to assist you with the entire process or any part of the mapping process as required.

Mapping your lubrication state is the simplest and most efficient way to achieving the 4Rs of lubrication.

TIP √ Consider starting your lubrication mapping process developing a lubricated machine register from the work order system. This can be built in spreadsheet format. Each lubricated machine can be broken down by lubrication type(s) employed (one machine may have more than one lubrication system and use both oil(s) and grease). Then the current lubricants used can be also be recorded.

This spreadsheet can now be used to build a list of lubricants in use in the plant top assist on the consolidation program and used to develop the machine maps and lubrication routes.

3

Lubricant Receiving and Storage Strategies and Tips

3.1 Lubricant Receiving and Storage Strategy One: Design your Lube Storage Facility with Foresight

Unfortunately, until a plant recognizes the true value of a managed lubrication program, lubrication projects are not always given the priority they deserve. Therefore, when a green light is given to actually implement lubrication related improvements, foresight should come into the decision process. This is especially true when designing a lubricant storage and handling facility. Points to consider in the design stage are as follows:

1. **Expansion plans:**

    *   If there are plans to expand operations within the next five years, it's crucial to incorporate flexibility and scalability into the design of the lubricant storage facility.
    *   Consider allocating space that allows for easy expansion or modification of the facility to accommodate additional lubricant storage needs as operations grow.
    *   Factor in potential increases in inventory levels and variations in product types that may occur with expansion.

2. **Real estate commitment:**

    *   Evaluate the available plant real estate and determine how much space can be dedicated to the lubricant storage facility.
    *   Balance the need for adequate storage capacity with other operational requirements and space constraints within the facility.

- Optimize the layout and design to make the most efficient use of the available real estate while ensuring accessibility and safety.

3. **Types of lubricants to store:**

- Identify the specific types of lubricants that will be stored in the facility, including oils, greases, fluids, etc.
- Consider the compatibility of different lubricants and ensure proper segregation and labeling to prevent contamination.
- Determine the required storage conditions for each type of lubricant, including temperature control, ventilation, and protection from moisture and sunlight.

Additional considerations in designing a lubricant storage facility may include:

- Implementing appropriate shelving, racks, or storage containers to organize and optimize space utilization.
- Installing proper ventilation systems to maintain air quality and prevent the buildup of fumes or vapors from volatile lubricants.
- Incorporating spill containment measures and secondary containment systems to mitigate environmental risks and ensure compliance with regulations.
- Implementing a robust inventory management system to track stock levels, monitor usage, and facilitate timely replenishment.
- Providing adequate lighting and security measures to ensure a safe working environment and prevent theft or unauthorized access.

By addressing these questions and considerations upfront, you can develop a comprehensive design for a lubricant storage facility that meets both current needs and allows for future growth and flexibility.

3.2 Lubricant Receiving and Storage Strategy Two: Implement a Cradle-to-Cradle (C2C) Lubricant Management Program

In a cradle-to-cradle lubricant management program a lubricant commences life as base oil refined from mineral crude or synthesized stocks, in which is then blended a variety of additives to make up a proprietary lubricant product ready for market delivery.

From the refiner/manufacturer the finished lubricant is transferred into bulk containers to continue its journey to the market supplier who offloads the bulk oil into their own storage tanks. The oil is then sold to the end user and delivered in pre-packaged pails and drums, or transferred once again into a bulk container for bulk delivery and transfer into the end user's bulk storage totes. At the end user's site, the lubricant is then stored and transferred into smaller containers used to fill up machine reservoirs.

If the lubricant additives are no longer effective, it is now changed out for new oil while the old oil is collected in dedicated used oil containers, classified, and stored ready to send to a re-refiner for recycling into marketable oil stocks once again as depicted in Figure 3.1.

Figure 3.1: A typical industrial oil journey/life cycle.

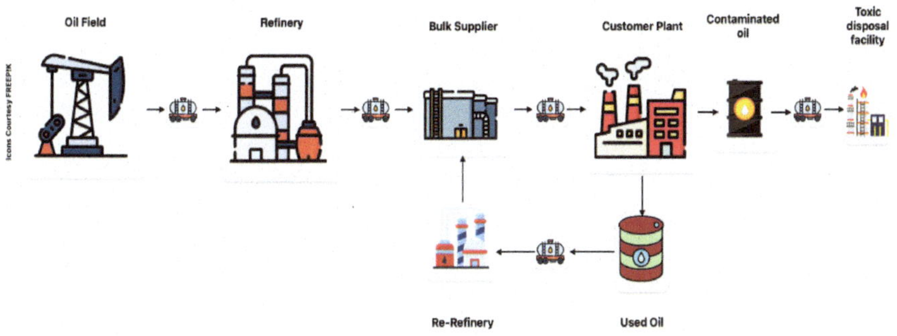

Each time a lubricant is stored or transferred it is at risk for solids and water contamination, both highly detrimental to bearings. As an end user, we gain control of lubricant cleanliness upon receipt from the supplier through an effective in-house lubricant storage and handling program. Prior to receipt, we can exercise a degree of cleanliness control through an audited cleanliness control lubricant purchase program set up in conjunction with the lubricant supplier(s).

Designing and building an end user world class lubricant management storage and handling facility requires the end user to view and treat lubricants with reverence using simple but effective management processes from the time lubricants are received, transferred, stored, dispensed, drained and collected for recycling. Adopting the following three-step approach the end user will have the tools needed to put in place the best program for their facility and budget.

Step 1. Understanding your lubrication requirements

Foundational to any lubrication management program is assuring the right lubricant choice is employed for the end user's machinery requirements and, more importantly, their operating conditions; the right lubricant being an engineered choice based on performance and economy.

Choosing the right lubricant is a specialized task best left to lubricant engineering experts. Fortunately, virtually all of the major lubricant manufact-urers and their suppliers offer the end user some form of a lubricant engineering service in which a lubricant engineer(s) performs a site visit(s) to determine the *least* number of lubricant types best suited to the end user's working environment and engineering requirements. This type of service is called a lubricant consolidation service or program and is a mandatory requirement of any lubrication management program.

Many consolidation audits are often offered as a no cost value added service in return for an exclusive lubricant supply contract. A consolidation exercise can considerably reduce costs by reducing the amounts of different products carried in stock, which saves on operating capital, storage and handling, and administration costs. A typical company can carry over 20 lubricant types in stock, which can often be reduced by 60% depending on the type of operation! Figure 3.2 depicts how NOT to store lubricants! Can you list the many things

Figure 3.2: How NOT to store lubricants!

Courtesy: **ENGTECH** Industries Inc.

that are wrong with this storage set up? See the end of the chapter for a list of wrongdoings.

Throughout the consolidation process, all unwanted and unused open and closed lubricant containers can be collected for recycling as only the consolidated lubricants are to be allowed on site once the program is activated.

Step 2. Designing and preparing a lubricant storage area

With a new consolidated lubricant list in place, the end user report should also contain information regarding the use of each lubricant relevant for designing adequate storage and handling that should include information on the following:

- Number of different lubricant products to be carried, stored and dispensed.
- Anticipated usage amounts for each lubricant—used to recommend the most economical purchase and storage container size. This data is also used to determine the real estate requirement for three-to-six-month supply of each lubricant and the type of filtering, dispensing and transfer equipment needed to ensure only contamination free lubricant ends up in the bearing surface area.
- In-plant geography of where each lubricant is used—used to determine economical lube routes and the logistical requirements (number of lubrication technicians required, lube truck, fork lift, lubricant delivery equipment, and for getting the right lubricant(s) to the right machine(s)).
- SDS sheets and staff training requirements.

With this information, the end user can now decide how best to design their lubricant storage and handling facility. Table 3.1 depicts the typical attributes of a storage and handling facility that must be factored into the design, a best practice and a reasonable alternative design are shown.

3.2.1 Location/size

As with any real estate, a good location is paramount; logistically important for receiving lubricants and dispatching them throughout the plant; larger facilities may also require controlled satellite locations. As equally important is protecting the virgin stock lubricants from the outside elements of extreme hot and cold temperatures (large temperature swings promote moisture condensation in the containers), wind and rain (bringing in lots of solid and water contamination possibility if the containers are not stored or handled with extreme care prior to opening for lubricant transfer).

Table 3.1: Typical attributes a storage and handling facility must factor into its design.

Attribute	Best practice design	Alternate design
Location	Indoors in a contained temperature-controlled environment	Outdoors in a contained environment protected from the elements
Size	Big enough to house 3–6 months supply of lubricants based on expected number and turnover rate of all lubricant products. Must be rated for fork lift truck use and big enough to store waste lubricants	Big enough to house 2–3 months supply of lubricants based on expected number and turnover rate of all lubricant products Must be rated for fork lift truck use and big enough to store waste lubricants
Ventilation	Motorized cross flow ventilation system c/w fresh air and exhaust ventilation/fan units	Prevailing wind vents and exhaust fan units
Fixtures	Storage containers and racking custom designed to suit the consolidation inventory flow, dispensing requirements and space limitations	Shelving designed to take stacked drum and pails containers
Transfer equipment/filtration	Dedicated filter transfer equipment for each lubricant type	Dedicated filter transfer equipment for each lubricant type
Spill control	In-floor central collection tank backed up with dry spill collection	Bermed or dyked areas with dry spill collection
Safety	Permanent plumbed eye wash station and portable eye wash station Post MSDS binders at entrance/exits	Portable eye wash station post MSDS binders at entrance
Stock control	FIFO Stock control	FIFO stock control
Waste oil control	Dedicated waste tanks for each lubricant product in the plant	Dedicated waste oil lay down area to store clearly labeled waste oil containers ready for recycling
Identification	Dedicate specific areas for each lubricant and waste oils and clearly label	Dedicate specific areas for each lubricant and waste oils and clearly label
Processes and procedures	Develop processes and procedure for sustainable operation of the facility and train all relevant staff	Develop processes and procedure for sustainable operation of the facility and train all relevant staff

All locations must be rated for and large enough to allow fork lift truck traffic. Indoor locations are to be temperature controlled, large enough to house 3–6 months of inventory and contain a plant exterior wall for a shipping and receiving dock door.

3.2.2 Ventilation

Because lubricants discharge vapors that might be harmful if allowed to accumulate, a good cross flow ventilation consisting of fresh air units and exhaust fans units is very important. Outdoor facilities may require their own cross flow ventilation if due to its location the building cannot take advantage of prevailing wind vents open to air (vents should be filtered using a furnace style filter) complimented with exhaust fan units.

3.2.3 Fixtures

Based on the lubricant turnover rate and storage container economics, storage and handling facilities may contain 300-gallon poly totes for bulk oil distribution, custom color-coded steel tank bulk oil storage and dispensing unit systems, drum racks designed to take palletized drums in the upright position or drum dispensing racks set on spill control platforms with the drum positioned on its side in the rack complete with a dispensing/metering valve system. Pails can also be used with the lesser used lubricants and stacked on pallets similar to the drums.

3.2.4 Transfer/filtration equipment

To move the lubricant to the machine, maintenance must transfer the lubricant from one container to another in the most non-contaminable way possible. For bulk containers, use of dedicated transfer/filter cart style dispensing units will ensure lubricant is moved from the bulk and pre-packaged supplier containers to the machine reservoir and/or dedicated closed pour containers similar to that shown in Figure 4.3 for transfer to the lube system reservoirs.

3.2.5 Spill control

Best practice is to slope storage room floors to a low point drain where spilled product can be collected into an accessible waste oil control tank. Local spills can be managed with dry spill absorbent products.

3.2.6 Safety

A permanently plumbed eye wash station is a must when dealing with petroleum-based products. Up to date SDS binders and standard operating procedures are to be posted at the entrance to the facility.

3.2.7 Stock control

Most lubricants are only rated for a shelf life between 6–12 months. Stock must be rotated on a regular basis following a FIFO (first in-first out) approach to stock control. Past due date lubricants should be returned to the supplier or recycled and the stock purchasing/usage history should be evaluated and adjusted accordingly.

3.2.8 Waste oil control

Mixed waste oils can be ten times more expensive to ship off site. Each lubricant used in the plant must have a dedicated collection tank/tote clearly marked for each waste lubricant type, the only exception being spilled oils. Used oil is likely to be transported to the storage and handling facility in a number of different containers that need to be disposed of according to local municipal and state requirements.

Oily rags can self-combust in a regular garbage can; collect them in a marked fireproof trash can.

3.2.9 Identification

Having gone to the trouble of consolidating your lubricants, clearly identify a dedicated area in the facility for each and every lubricant clearly label marking in large letters (2–3 inches high) the lubricant ID. Develop a plan drawing of the facility identifying locations of each lubricant and waste tanks and post at the facility entrance.

3.2.10 Processes and procedures

Best practices are not only rooted in the design but in a sustainable operation of the facility. Be sure to develop, map put, and train all staff on all processes and procedures related to use of the storage and handling facility

Step 3. The paperwork

To ensure your supplier provides you with the cleanest lubricant product, follow these simple rules before accepting bulk lubricant deliveries in your new storage and handling facility.

- Insist on receiving the lubricant Certificate of Analysis (COA) for each lubricant delivered and keep this document on file until the batch of lubricant has been used.
- Never assume all lubricants are delivered as per their COA document specification.
- Set up a delivery acceptance agreement with the supplier to deliver lubricant based on the COA and/or a set of internal minimum cleanliness (see Table 4.1 for ISO 4406:1999 guidelines) and viscosity specifications (within ±10% of the COA specification).
- Establish an oil quality analysis test acceptable to end user and supplier, and a develop a service level agreement that outlines the lubricant condemning levels and remedial action requirements should the lubricant fail the quality test on delivery.
- Perform quality testing regularly taking a bulk sample after the tanker truck lines have been flushed prior to transfer, and from the center of any supplier pre-filled containers.

Once in place, the end user must then perform a series of updates to manage the new lubricants, which will include:

- Update the asset management PM system to reflect the new lubricant choices on the work order.
- Update machinery bill of materials.
- Update the inventory portion of the asset management system to reflect the new products.
- If applicable, update lubricant labeling on equipment reservoirs.
- Update SDS manuals.
- Update equipment manuals.
- Update purchasing records.

TIP √ Don't be afraid to invite/engage your lubricant supplier to assist in formalizing your paperwork requirement. Chances are they already have clients who have already put in place a managed program they can draw from to assist you in your program development.

TIP √ Institute a partnership agreement with your lubricant supplier(s) that clearly outlines each partner's responsibilities. This agreement will also spell out the specific acceptance requirement to be outlined on the COA, and the responsibilities and actions of each party if the lubricant does not meet the delivery cleanliness requirements.

3.2.11 Figure 3.2 Wrong-doings

Figure 3.2 shows a lubrication storage area complete with an oil catch container and an eye wash station in place. There is also a fire hose station c/w extinguishers and a wall locker for SDS sheets and labelling materials. The following is a list of poorly executed lubrication management and safety elements that require immediate remediation.

1. The lubricant oil catch pan is placed directly in front of the fire hose cabinet and fire extinguishers. The stored lubricant containers are automatically a safety hazard as they directly block access to the fire-fighting equipment.
2. A large packing container is blocking access to the eye wash station and is an immediate safety hazard.
3. The lubricant containers are poorly labelled. It is difficult to tell if the pails contain virgin stock oil or used oil.
4. There are packs of poorly marked chemicals poorly placed on a lubricant container. There is no fast way of knowing if these two substances are incompatible. This is automatically a safety hazard.
5. The containers are not stored in such a manner as to facilitate a LIFO stock rotation.
6. The wall cabinet is open and represents a physical hazard to any person working in that area.
7. There is spilled lubricant on the floor to the right side of the oil catch pan
8. There is no spill kit close by.

3.3 Lubricant Receiving and Storage Strategy Three: Choose the Correct Lubricant Storage Reservoir for the Job

The reservoir, according to Webster's dictionary, serves as a liquid storage device, especially for liquids like lubricants or hydraulic fluids. Among these, two main categories stand out: storage reservoirs and working reservoirs, also known as production reservoirs.

3.3.1 Storage reservoirs

In their simplest form, lubricant reservoirs act as storage vessels, transporting and holding lubricants for future transfer into smaller working reservoirs. For instance, tanker trucks deliver oils to sites, where they are then transferred into on-site storage reservoirs, commonly 300-gallon polyethylene totes. These totes are favored for their rust-free nature and easy visual inspection due to their transparency (see Figure 3.3 below). To prevent bulging when full, they're often encased in steel frames for rigidity and stacking convenience.

Figure 3.3: Double plastic tote storage system with steel cage racking.

Courtesy: **ENGTECH** Industries Inc.

Grease storage differs slightly, with greases often arriving in one-time-use drums or pails that can be easily transported and their contents manually transferred to the working grease lube system reservoir at the machine. Grease cartridges serve as mini reservoirs, transforming grease guns into pumping systems. Typically, storage reservoirs are basic vessels with lids or large openings for bulk filling. They may feature drain ports for moisture removal and tap valves for dispensing lubricant.

Since storage reservoirs are seldom filled to the brim, a "head space" is created, prone to accumulating contaminants from ambient conditions or open filling. To counter this, desiccant-style breathers are recommended, allowing airflow while capturing moisture and solids.

Never store new oil containers outdoors without protection from the weather. If possible, strive to store all oils and greases in a dry indoor location at an acceptable temperature range between 0 F and 110 F ($-18°$C and $43°$C).

If using steel drum storage containers, ensure all bungs, lids and breathers are always in place.

3.3.2 Working reservoirs

Unlike their storage counterparts, working reservoirs are integral components of machines or systems, often requiring unique design shapes to fit specific spaces (think of how every gas tank reservoir on a car is a different shape). In closed-loop oil systems, their sizing is critical, balancing lubrication needs with pump and piping sizes. These reservoirs play a vital role in recirculating oil systems, cooling oil before it's pumped back to bearings. Internal baffles aid in this process, facilitating slow oil movement across the reservoir to dissipate heat.

Working reservoirs typically incorporate suction filters and external indicators for monitoring oil levels. For automatic filling, low- and high-limit sensors coupled with solenoid fill valves are employed. Similar to storage reservoirs, desiccant breathers help maintain cleanliness and head space moisture control.

In terms of maintenance, reservoirs require regular attention to prevent contamination and ensure efficient operation. This includes:

- Cleaning reservoir exterior surfaces so they don't insulate and heat up lubricant inside the reservoir, or present an opportunity for dirt to contaminate the lubricant product when open filling the reservoir.
- Monitoring oil levels daily.
- Inspect/clean/replace filters and breathers regularly.
- Employing off-line bypass cleaning methods to maintain oil quality.

In summary, lubricant reservoirs serve as crucial components in industrial settings, requiring careful design and maintenance to uphold system reliability and performance.

TIP √ Replace old style steel breathers with desiccant style breathers that change color when the breather requires replacement, as seen in Figure 3.4. Note: for more information on desiccant breathers refer to Chapter 7.

Figure 3.4: Old style breather versus new style desiccant breather.

Courtesy: ENGTECH Industries Inc. and Des-Case Corporation

TIP √ On large reservoirs for continuous systems, consider installing a duplex primary filter and valve arrangement so that the lubricant can be routed through the clean filter while the dirty filter can be isolated and changed ready for the next exchange cycle.

4

Lubricant Application and Operational Strategies and Tips

4.1 Lubricant Application and Operational Strategy One: Set Up your Single Point Lubricator to Immediately Function when Installed

Utilizing programmable single point lubricators (SPLs) is a common practice among many industries to maintain the efficiency of equipment bearings, especially those in hard-to-reach locations that require regular lubrication. Despite their convenience, users often encounter issues where these units fail to dispense lubricant, even when fully charged and displaying active status.

SPL devices typically operate with electro-mechanical discharge pumps or electro-chemical reaction chambers to continuously deliver grease to a single lubrication point for up to two years on a single charge. However, several factors can cause the unit to appear functional but fail to deliver lubricant effectively:

1. Blocked or collapsed lube line: This is where dirt or debris causes a partial blockage or collapse of the lube line that in turn creates line back pressure, stalling the unit's operation. In addition, the unit "stand-off" line may have been crushed or bent by a fork lift or accidental movement.
2. Bearing misalignment: If the bearing turns within its housing, it can block the lubricant entrance to the bearing raceway, hindering grease delivery creating system back pressure restricting lubricant flow.
3. Cold weather operation: In cold weather when using #2 or heavier greases, the lubricant can solidify, causing the unit delivery to stall.

4.   Incorrect setup: Setting the unit on the longest delivery settings without proper calibration can lead to operational and delivery issues.

To address these challenges, it's essential to adhere to the following guidelines:

1.   Follow manufacturer's instructions: Always consult the manufacturer's instructions before using an SPL unit to ensure correct setup and operation.
2.   Pre-install set up on shortest delivery cycle: Run the SPL unit on its shortest delivery cycle for an extended period, typically 12 hours or more, before installation to verify its functionality.
3.   Use appropriate grease thickness rating: Select the appropriate NLGI numbered grease for the operating temperature to ensure optimal performance. Units can be pre-purchased with grease of your specification.
4.   Inspect delivery line integrity prior to installation, inspect the delivery line to ensure it's free from damage or obstructions that could impede grease flow.

Figure 4.1: Typical single point lubricator (SPL) with install date clearly marked.

Courtesy: **ENGTECH** Industries Inc.

By following these recommendations, users can mitigate common issues associated with SPL units and maintain their equipment's lubrication effectiveness.

TIP √ If installing a new delivery line, ALWAYS prefill the line prior to installing the SPL lubricator.

TIP √ Be aware of your plant elevation. High altitude installations will dispense at a faster rate, and visa versa. Check altitude adjustment requirements with manufacturer recommendations.

TIP √ A sure visual sign of unit back pressure from line blockage is a dislodged SPL unit's screwed on plastic top cap.

TIP √ At time of installation, mark the SPL unit body with the date of installation and delivery setting (see Figure 4.1). Also, if the unit uses an opaque plastic body, the lube level can be seen and a line drawn at the level. This line level can be used for future inspections to confirm the lubricant is being pumped into the bearing.

4.2 Lubricant Application and Operational Strategy Two: Regularly Check Temperatures of Working Lubricant Reservoirs

To design and build a reliable gearbox, the gearbox designer must consider many factors. Arguably, the most critical being the size of the gearbox, as it must not only fit within the physical confines of the machine footprint, it must also be accessible for filling and cleanout. For total loss systems, the reservoir must be large enough to hold one to two weeks amount of lubricant before the need to refill. If it serves a continuous lubrication system, the reservoir should be large enough to hold at least two to three times the system volume and accommodate cooling baffles. These serve to cool the lubricant sufficiently before it is sent back to the bearings where once again the lubricant can lubricate and purposely absorb the bearing heat.

When designed correctly, the reservoir temperature should be warm to touch but cool enough that you can leave your hand on the reservoir comfortably. When measuring with an I/R thermometer or camera, similar to those depicted in Figure 4.2, the temperature should be around 40°C or 104 F.

Simply put, cooler bearings equal longer life. According to the Arrhenius rule—a temperature change-dependent failure rate rule—for every 10°C (18 F)

Figure 4.2: Raytek I/R thermometer and a Fluke hand held I/R camera.

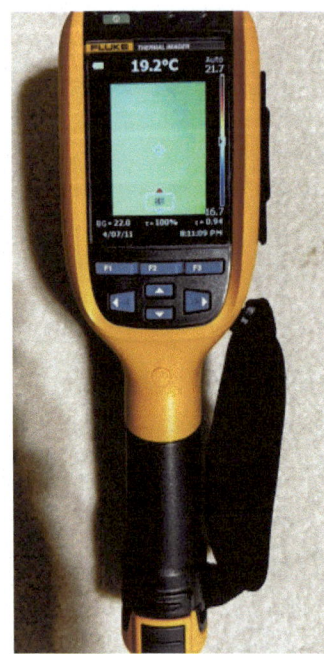

Courtesy: **ENGTECH** Industries Inc.

rise in temperature of the oil, the lubricant lifecycle is halved, the inverse also being true.

Should you find your gearbox is running at temperatures somewhat higher than 40°C or 104 F, (or you are unable to comfortably hold your hand on the gearbox) the following recommendations will give you some starting points for determining a root cause and remedy for the overheating situation:

- Check oil level and for leaks: Ensure the gearbox has sufficient oil and there are no leaks that could result in inadequate lubrication and cooling.
- Assess oil condition: If the oil is old, it could be oxidized, leading to sludge buildup and increased viscosity. Ensure that the oil is changed regularly with compatible gear oil of the correct viscosity.
- Inspect pressurized system: Check for any blockages in the system that could be impeding oil flow and cooling.

- Clean the gearbox: Remove any debris or dirt from both the internal and external surfaces of the gearbox. Ensure that the oil fill cap and reservoir breather are in place to prevent the buildup of a thermal blanket.
- Check for external heat sources: Investigate if there are any new external heat sources or changes in operating parameters that could be contributing to the increase in gearbox temperature. Use an infrared thermometer or camera to identify hotspots.

It's important to note that there could be multiple factors contributing to the overheating issue, so a comprehensive inspection and troubleshooting process may be necessary. If these recommendations do not resolve the problem, it's advisable to seek the assistance of a professional lubrication management consultant who can provide further expertise and guidance.

TIP √ Invest in an infra-red thermometer or infra-red camera system to take regular temperature checks (see Figure 3.7).

TIP √ Use a 105 F and a 120 F temperature crayon to invisibly write 105 with the 105 F crayon and 120 with the 120 F crayon on the reservoir. If those temperatures are reached, the marking will become visible to the operator who can them bring it to the supervisor's attention for further maintenance action.

4.3 Lubricant Application and Operational Strategy Three: Use Dedicated Filter Carts to Clean and Transfer in Plant Oil Lubricants

Controlling contamination is a critical aspect of effective lubrication management, particularly concerning lubricant cleanliness. Clean oil is not only essential for extending the life of lubricants but also for maximizing the lifespan of machine components, particularly bearings.

Portable filter carts serve as efficient and practical tools for ensuring lubricant cleanliness in various industrial environments. They are versatile and find application in several key areas within a lubrication-management program, consider the following examples:

- **Oil transfer**: Portable filter carts facilitate the transfer of oil from its original container into the reservoir of the machine. This ensures that the oil introduced into the system is clean and free from contaminants that could potentially damage the machinery.
- **Pre-filtering and cleanup of new oil**: Before introducing new oil into a machine, it's crucial to pre-filter and clean it to remove any impurities or contaminants that may have been present during storage or transportation. Portable filter carts are employed for this purpose to ensure that the virgin stock oil meets the required cleanliness standards before being used in the equipment.

- **Reconditioning and cleanup of in-service oil**: Over time, oil in service can accumulate contaminants such as dirt, debris, and moisture, which can degrade its quality and effectiveness. Portable filter carts are utilized to recondition and clean the oil while it's still in use, extending its lifespan and maintaining optimal performance of the machinery.

Additionally, specialized filters can be installed on the outlet side of the filter cart to specifically target and extract free and emulsified water present in the oil. This is crucial for preventing water-related issues such as corrosion, oxidation, and reduced lubricating properties. Figure 4.3 shows a typical dual-filter cart.

Overall, the use of portable filter carts as part of a comprehensive lubrication-management program significantly contributes to maintaining lubricant cleanliness, prolonging equipment life, and minimizing downtime due to lubrication-related issues. The ability to control contamination is an important aspect of any lubrication-management program, especially where

Figure 4.3: Typical dual filter style cart.

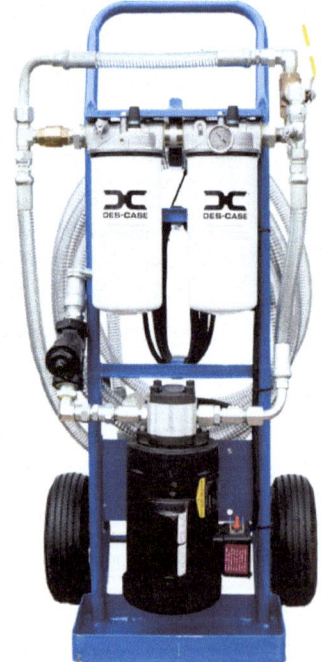

Courtesy: Des-Case Corporation

lubricant cleanliness is concerned. A constant supply of clean oil is essential to lubricant life and, more importantly, bearing life.

In addition, use of specialized filters on the outlet side can extract any free and emulsified water present in the oil.

4.3.1 Functionality

The primary function of any filter cart is to filter fluids. A typical cart design will employ a two-stage filtration approach in which a gear pump is connected to both filters. The inlet, or suction, side is the first-stage, low-pressure side (approximately 5 psid) designed to capture larger contaminant particles exceeding 150 microns in size.

Oil is pumped through the inlet filter to the second-stage, high-pressure (approximately 25 psid) outlet (or delivery side) filter designed to capture much smaller particulate matter that can be filtered to less than 5 microns in size, depending on the filter rating used.

4.3.2 How clean is clean oil?

Oil cleanliness is universally measured using the ISO 4406 cleanliness code rating system. This is a standard that quantifies the number of contaminant particles, 4, 6, and 14 microns in size, that are present in a 1-ml lubricant sample, and compares them with a particle concentration range, resulting in an ISO-range number value. Compare the various samples and cleanliness levels in Table 4.1.

The number of passes through the filter cart to achieve the appropriate cleanliness level will depend on the "start" and "finish: cleanliness level and the filter types and rating in use. Oil analysis will be required to establish cleanliness levels. Choosing a suitable combination of pump and filter size/type will require consultation with the filter-cart manufacturer who will need to understand your working environment and type/viscosity of oil(s) you use.

The rate of cleanup (speed) will depend on the reservoir size, pump flow rate, and the cleanliness-rating delta. What can be measured immediately is the time to perform one complete filter pass through the cart, as calculated using the following formula:

(Reservoir size $\times$ 7)/filter cart flow rate = time for a single pass filtration

Table 4.1: ISO Lubricant sample cleanliness rating table.

ISO Range	Description	Application	Life extension	Dirt p.a.*
22/20/17	**Very dirty**	• Unsuitable for any lubrication system (H,L,G)	0.5× (H) 0.75× (G)	600 kg
19/17/14	**New oil**	• Medium/low pressure systems (H,L) • Non-critical gear systems (G)	1	140 kg
17/15/12	**Light contamination**	• Standard hydraulic and lube oil systems (H,L) • Standard gear system (G)	1.6× (H) 1.3× (G)	34 kg
16/14/11	**Clean**	• Servos and high-pressure hydraulics (H,L) • Critical gear systems (G)	3× (H) 1.7× (G)	17 kg
<14/12/10	**Very clean**	• All oil systems (H,L,G)	4(H)–2×(G)	8.5 kg

H: Hydraulic oil, L: lube oil, G: gear oil.
* Dirt through pump at 200 L/min, 230 work day, single shift in 1 year.

Courtesy: **ENGTECH** Industries Inc.

Example for 60 gal reservoir:

60 gal × 7/10 gpm = 42 min for a single-pass filtration (1 × filtration of reservoir capacity).

If the plant's lubricants are consolidated and cleanliness levels are known, a matrix can be developed to determine how many passes are required to filter to acceptable cleanliness levels.

Work with the filter cart supplier to determine the right pump and filter choice for your plant requirements.

TIP √ To eliminate cross contamination of lubricants, each filter cart must be dedicated to a single lubricant use for transfer and cleaning of lubricants. Pilot the filter cart program with the most-critical and/or most-utilized plant-lubricant type.

TIP √ Always clean the unit after each successful transfer operation, paying particular attention to the wand ends and open drip tray under the filters and pump area. Open oil is a dirt attractant and can be transferred unwittingly if the cart and its components are not kept scrupulously clean.

TIP √ Unless specified, a typical filter cart is sold with open-end transfer wands fitted to the delivery and suction hose ends designed to slide easily into the reservoir openings of the donor and recipient reservoirs. In a program designed to filter contaminants from the oil, this type of delivery fitting can allow moisture and dirt contamination into the respective reservoirs during the transfer process. To combat this, and ensure a contamination-free transfer process, fit the filter cart delivery/return hose ends and reservoir fill/drain ports with quick-lock-style couplings. As the reservoir is now airtight, it will also require a quality desiccant-style breather to be fitted and, in the case of larger capacity reservoir, a closed-loop expansion tank.

TIP √ Specify kink-resistant flexible suction and delivery hose to prevent pump cavitation. Clear hoses allow a visual reference of the oil flowing through the lines.

TIP √ The cart's electric motor will require access to electricity. Ensure that an electrical outlet is within easy reach of the unit's electrical cord. If the cord is short in length, consider mounting a retractable electrical cord caddy on the unit with enough cord length to reach the nearest electrical outlet.

TIP √ Paint a lined box similar to a lay-down area as close as possible to the oil reservoir that's to be serviced. This allows a cart to be positioned and used quickly without obstruction, and within reach of its hose and wand assemblies.

TIP √ Place the cart on a preventive-maintenance (PM) check program prior to every use to ensure the unit's filters don't go into bypass mode from being too dirty.

4.4 Lubricant Application and Operational Strategy Four: Don't Believe Everything you See on the Internet

When little or no formal training has taken place, a favourite "go-to" place is the internet, where both knowledgeable and "not-so" knowledgeable people gather to dispense their wisdom for free. When it comes to lubrication practices, information gleaned from online forums and posted videos could cost your operations dearly.

When performing a personal test internet knowledge, I visited a representative sample of "shade-tree" lubrication learning sites offering free advice and training; unfortunately, the result was worrisome.

While certain elements of each video's proffered advice could be construed as useful to viewers (or "trainees"), many of the shared practices could easily prove to be harmful to a bearing or the person performing the lubrication delivery method being taught.

Some of the highlights, or should I say "low-lights" included:

- Use of a 50-year-old grease gun that transferred rust and contaminants into a new grease tube that will eventually make its way into a bearing.

TIP √ Discard the relic and start again with a new gun.

- Cross-contaminating grease by replenishing an empty delivery tube with a totally different type of grease product, without first cleaning the gun thoroughly or purging all old grease from the tube.

TIP √ Grease guns are inexpensive. Always use dedicated grease guns for each grease in the plant.

- Using an open-flame torch to "liquefy" and free old hardened grease, with no nod to safety equipment or safe practices.

TIP √ Open flame use on petroleum products is a sure-fire risk that can accelerate rapidly. For this type of problem, it is better to use a hot air heat gun used to strip paint.

- Incorrectly disassembling a grease gun to replace the delivery tube, creating a greasy mess and introducing unwanted air and contamination into the pump.

TIP √ Learn grease gun anatomy and how to disassemble and reassemble a grease gun correctly.

- Spinning a dry, unlubricated bearing and handling it with bare hands during an over-packing procedure.

TIP √ Put in place a policy for storing and handling bearings. Bearings are manufactured in a HEPA filtered "white" room environment then wrapped for protection; most are not lubricated. Touching the bearing surfaces with skin oils can reduce the life of the bearing significantly as can spinning the bearing dry and unlubricated. Handle with lint free gloves.

- Dropping a new bearing into metal shavings on the floor, then cleaning it "as good as new" with compressed air before greasing and installing.

TIP √ Put in place a policy for storing and handling bearings. Scrap all dropped bearings onto dirty surfaces. Realize that bearings dropped in contaminated areas can never be fully cleaned. Micron size contaminants immediately act as two and three body abrasives that will score the bearing surfaces causing it to prematurely fail. The cost of a new bearing pales in comparison to the cost of a premature bearing failure.

- Filling a bearing with so much grease and pressure causing the bearing seal to eject from the bearing

TIP √ A bearing need only be filled to 30–50% of the bearing cavity area. Provide lubrication training and certification to all persons handling a grease gun.

By all means, *enjoy* as many online lubrication-related forums and videos as you wish. But before that happens provide your lubrication team with formal training and certification from a reputable lubrication expert or qualified certification organization like the ICML or STLE. Then internet videos can be watched as a training exercise to spot problem advice versus quality advice. Remember: You almost always get what you pay for!

TIP √ Get certified!

4.5 Lubricant Application and Operational Strategy Five: Extend your Lubricant's Life

Every lubricant is an engineered product. If a lubricant is to lead a long productive and healthy life, it requires your assistance to combat four main adversaries. These being: (1) dirt, (2) heat, (3) moisture, and (4) apathy.

Dirt clogs filters, creates sludge and destroys machined surfaces. Heat directly affects the oxidative (useful) life of the lubricant (for every 18 F temperature raise, oil life expectancy is halved). Moisture (water) attacks the base oil and prematurely strips out the additive package causing sludge and oxidation. Not caring or understanding the need for an engineered lubrication program exacerbates the problem significantly.

Adopting the following simple tips can significantly extend your lubricant's life, save significant bottom line dollars and boost machine reliability.

TIP √ Work with your lubricant supplier to perform a *lubricant consolidation program* to reduce and optimize the number of lubricant SKUs used in the plant. This will reduce the risk of in-service cross-contamination, reduce your storage and handling requirements, ensure your lubricants are always fresh.

TIP √ Introduce professional grade dedicated transfer and delivery equipment designed to combat lubricant cross-contamination and dirt contamination during transfer.

TIP √ Introduce a *machine cleanliness initiative* to quickly identify leaks, moisture invasion, and prevent dirt from contaminating the lubricant and becoming a thermal blanket raising the lubricant temperature, all of which significantly reduce the lubricant life cycle.

TIP √ Utilize a *4R lubrication initiative* to tune up your lubrication delivery methods and systems to ensure the *R*ight lubricant is placed in the *R*ight amount, in the *R*ight place, at the *R*ight time,

TIP √ Introduce a *condition-based oil analysis program* to determine when lubricants should be deep cleaned, reconstituted, or changed out. Don't change your oil too early or too late!

4.6 Lubricant Application and Operational Strategy Six: Insure Against Catastrophic Failure due to Lubricant Incompatibility when Changing Over to a New Lubricant

When no structured lubrication program is in place, the mixing of lubricants—both greases and oils—can be endemic, and is a major cause of lubricant failure due to cross contamination of base oils and/or additive packages. A lubricant containing acidic additives added to a lubricant containing base or alkaline additives can very quickly "neutralize" a lubricant's effectiveness and protection ability and result in catastrophic lubricant and bearing failure.

Similarly, when converting a machine over to a new lubricant, depending on the lubricants' compatibility, you may or may not be required to flush the old lubricant out of the reservoir and bearings with a neutral flushing oil prior to filling and using the new lubricant.

Before proceeding with any lubricant "swap" **ALWAYS** consult with the new, or preferred, lubricant manufacturer to establish if they have already tested compatibility with your old lubricant, or are willing to test compatibility on your behalf.

If no information is available or forth coming, and you are unable to establish compatibility, consider performing your own compatibility testing as follows:

1. Take virgin samples of both the old and new oils and blend three mix samples of both lubricants in a 50:50, 90:10, and 10:90 blend ratio.
2. Send all nine blended samples to an oil analysis laboratory and have them test for filterability, sediment, and color/clarity. In addition, have the lab perform a RPVOT (rotating pressure vessel oxidization test to determine an oil's resistance to oxidation) and a storage stability comparison.
3. If the lubricated asset is a critical machine perform steps 1 and 2 three times and instruct the laboratory to normalize the results for each ratio.
4. Ask the laboratory to determine if there is any cross-contamination risk in their final report.

TIP √ Because an RPVOT test is expensive to perform, you may choose to have the new lubricant manufacturer recommend a neutral flushing oil and forego testing if there are not that many lubricant changeovers to complete.

When dealing with greases, a similar process is followed in which two samples are mixed in a 75:25, and 25:75 ratio and sent to the lab to test for consistency, dropping point, and shear stability.

Understanding lubricant compatibility prior to any lubricant changeover will insure against any lubricant incompatibility failure shortly after changeover.

4.7 Lubricant Application and Operational Strategy Seven: Control Grease Application Consistency with Grease Gun Standardization

Not all grease guns are created equal and in the hands of the untrained, a grease gun must be considered a lethal weapon!

There is no standard grease gun, they may all look similar but displacement and hydraulic pressure rating vary greatly from gun to gun. For example, a manufacturer may offer two lever arm actuated grease gun models; grease gun

Table 4.2: Grease gun comparison table.

Grease Gun	ID #	Capacity	Delivery	Pressure
	F104	3oz bulk	1oz-25 stroke	**2500psi**
	525	14oz cartridge	1oz-33 stroke 1oz-60 stroke	**4500psi** **10,000psi**
	4015	14oz cartridge	1oz-7 stroke	**1700psi**
	6268	12oz bulk	1oz-24 stroke	**15,000psi**

Images Courtesy Alemite Corporation

Courtesy: **ENGTECH** Industries Inc

number one is a low-pressure gun rated to deliver 1 fluid ounce of grease in 7 strokes (shots) at a pressure of 1700 psi. Grease gun number two is a similar looking sister product rated to deliver 1 fluid ounce of grease in 24 strokes at a staggering 15,000 psi! To add to the confusion, the same manufacturer may offer a pistol grip actuated two-stage grease gun that delivers one fluid ounce of grease in either 33 strokes or 60 strokes at a corresponding pressure of 4500 psi and 10,000 psi, see Table 4.2.

To further confuse matters, many manufacturers do not state the grease gun delivery shot volume or weight, or maximum delivery pressure on their gun or in their literature—you will need to ask them directly or purchase a better-quality gun that does state their operating specifications.

If a work order job task asks for four shots of grease, the actual amount delivered will vary greatly depending on the grease gun used and how it is set up. Over-lubrication is a huge issue in manual greasing that causes bearing temperature spikes due to fluid friction; the issue is magnified further when a PM task simply states, "grease as necessary". This simple instruction gives no repeatable, definitive direction to the grease gun operator regarding the amount of grease required and, as a result, many bearings are not only overfilled, but can also lose their seals under the tremendous internal hydraulic pressure created by the grease gun when the bearing cavity is full, allowing dirt and debris to eventually be pulled into the bearing cavity.

4.7.1 Implementing an engineered manual lubrication program

Individual bearing lubrication requirements are easily calculated and ideally a bearing cavity only needs filling to approximately 30% to 40% volume. If a single point manual grease gun lubrication program is your chosen lubricant delivery method, the following recommendations and tips will help standardize your approach and significantly reduce premature bearing failures due to ineffective lubrication practices:

1. Perform a lubricating grease consolidation program to reduce the amount of grease types in use to lower chances of cross lubricant contamination.
2. Collect and remove all current grease guns from the plant and replace with a good quality single model grease gun.
   TIP √ Donate the old grease guns to maintainers and operators to take home for personal use
3. Perform a grease gun displacement check.
   TIP √ Pump 10 strokes or shots of grease into a large calibrated syringe then read off the number of cubic centimeters or cubic inches volume and divide by 10 to get the actual volume displacement per shot or stroke.
4. Calculate bearing requirements and mark on a bearing schematic that can be attached to the machine or printed with the PM work order.
   TIP √ If machine and lubrication mapping has not taken place, take a photograph showing each bearing location and mark the photo up in photoshop with a bearing number I.D. number and number of grease shots required.
5. Optional: Tag individual grease points with a color tag to denote the grease type and markup the number of lubrication shots required per PM schedule.
6. Perform grease gun operator training.

4.8 Lubricant Application and Operational Strategy Eight: Always Pre-fill Grease Lubricant Lines when Installing New System Lines or Replacing Damaged Lines

A typical automated or semi-automated grease lubrication system consists of a pump assembly connected to a number of distribution blocks or points depending on the system type.

Both types of lubrication system must be treated as a high-pressure hydraulic system for oil or grease delivery. As such they require pre-filled lines prior to start up so that the small apportioned amounts of grease discharged at the distribution point or multipoint block can simultaneously hydraulically push an equal amount of grease or oil through the secondary line into the bearing.

Filling the main feeder lines to the distribution points or blocks can be achieved in minutes by actuating the lubrication pump a number of cycles until oil or grease is witnessed at the distributor valve or block. However, using the same lube pump to fill the secondary lines from the distributor valve or block is not impossible but can take a very long time due to the apportioning aspect of the lubricant in the secondary side of the system.

4.8.1 Pre-filling a single/dual line system type

A single line and dual line injection system design allows the main line to be piped very close to the bearing point as the single discharge distributor valves are connected directly into the main line in a chain formation. This allows the secondary lines to be very short and easy to fill manually with a grease gun prior to attachment to the distributor and the bearing.

4.8.2 Pre-filling a progressive divider system type

Unfortunately, this is not the case with progressive divider style systems that have relatively short main lines connected to a centrally located distributor block, which are then connected with much longer secondary lines to the individual bearing point(s). This system type requires a simple multi-step process to ensure all air is expelled out of the system and all secondary lines are filled ready for immediate start-up and use.

Step 1. Virtually all progressive divider blocks secondary discharge points have interconnected discharge points at both front and sides of the divider block. As only one discharge point is used in service, the unused point (usually the front) is plugged. Once all secondary lines are connected to the block the bearing connection end of the lines remain unconnected; see Figure 4.4.

Step 2. Charge the main line of the system with lubricant by activating the lubrication pump until the main line connecting the pump to the divider block is full.

Step 3. Undo the interconnected plug at the first secondary line being filled and install a regular grease nipple.

Figure 4.4: Progressive divider block with overpressure indicators and plugs in the secondary discharge ports.

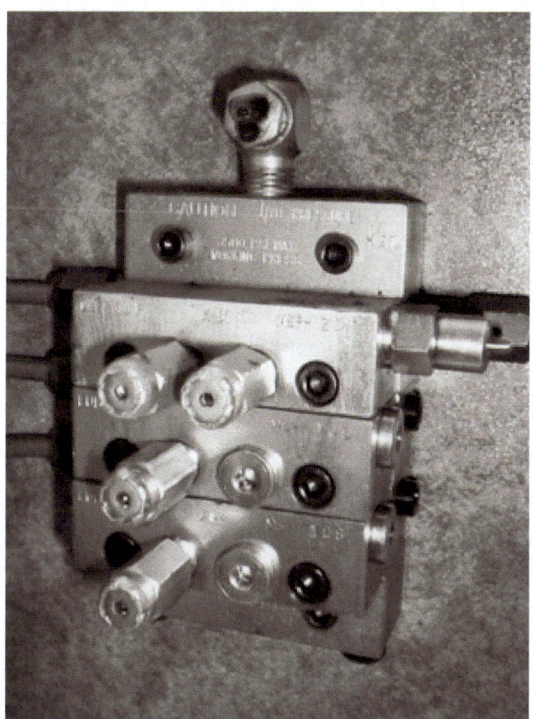

Courtesy: **ENGTECH** Industries Inc.

Step 4. Connect a manual or automated grease gun and using the exact same grease as in the system reservoir hand fill the line until grease appears at the corresponding bearing point end.

Step 5. Connect the grease filled secondary line to the bearing point and clean off with a lint free rag.

Step 6. Remove the grease nipple from the block and re-plug the block.

Step 7. Repeat for all secondary delivery lines and you are now able to place the lubrication system in operation.

4.9 Lubricant Application and Operational Strategy Nine: Choose your Lubricant Delivery Lines Wisely

When specifying a lubrication delivery system, much time is spent specifying the pump, its control system, and the distribution metering devices. Little time is afforded to specifying the distribution lines delivering the lubricant to the bearing. These are the weakest components in the lubrication system as they have to be shaped, bent and attached to the machine without stress or damage.

Lubrication lines are available in numerous materials that can have an effect on price, longevity, and reliability.

4.9.1 Copper lines

Pure copper lines have long been a favourite line material over the years. One of copper's perceived benefits is the softness of the metal that makes it a very easy and fast material to bend, shape and install. This can also be a drawback as it is also easy to work harden the material during the initial bending process causing stress cracking of the line surface.

Line hardening can accelerate as a result of any subsequent machine vibration, all of which releases a fine copper "dust" into the fluid. As with any tubing or piping material used to carry fluid, erosion wear caused by the fluid moving over the tubing surface (think of smooth rocks in a stream or river bed)—especially at sharp bends or corners—takes place and releases copper particulate into the fluid over time.

In a recirculating system the trace copper will attack the antioxidant packages present in many industrial and automotive oils. Interestingly, copper is the primary material used in laboratory testing to accelerate oxidation when testing lubricants!

In areas where large temperature swings take place moisture can build up in the lines and release copper ions. This is important in refineries as distillate fuels and oils can contain sulfur-based compounds called mercaptides. Copper and brass are not compatible with mercaptides and together form a copper gel that is similar to an insoluble grease that will plug filters and lines. Additives to prevent this from occurring are available.

4.9.2 Nylon/plastic lines

Nylon or plastic lines have a high degree of flexibility and are often employed in systems where vibration is an issue, or where a lubricated bearing is located as part of a moving or oscillating piece in the machine. These lines can be inexpensively and quickly installed in a bundled harness fashion (Figure 4.5).

The main issues with nylon/plastic lines are their low-pressure capacity when compared to steel lines. When a problem does occur, troubleshooting and reparation requires a lot of effort to isolate the line from the "bundle". The system will aways have a semi-permanent look to it and as such does not always receive the attention it deserves.

Figure 4.5: Progressive divider block with bundled nylon secondary lines.

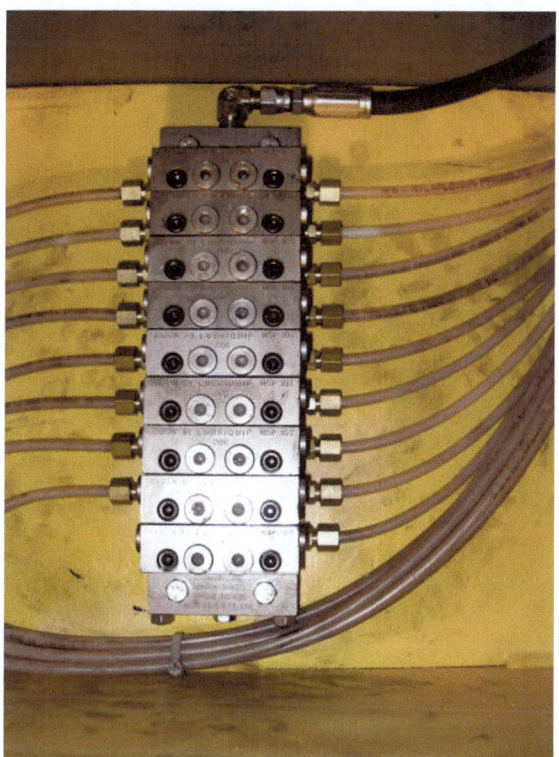

Courtesy: **ENGTECH** Industries Inc.

Figure 4.6: Progressive divider block with steel secondary lines.

Courtesy: **ENGTECH** Industries Inc.

4.9.3 Steel lines

Steel lines are by far the most robust and permanent solution for any lubrication system and can perform reliably in most shock and vibration environments. Steel lines can be bent easily into tight radius curves using a small handheld tube bender and will always look like a professional installation (Figure 4.6).

Where there is machine movement requiring a line flexibility, use of a matching inside diameter quality hose with steel crimped ends is recommended.

TIP √ When choosing a premium lubricant line material for damp the ideal choice is copper coated (on the outside) steel.

4.10 Lubricant Application and Operational Strategy Ten: Consolidate your Lubricant Choices

An integral component of any robust lubrication management regimen involves executing a lubricant consolidation process that meticulously assesses and documents all existing lubricant products onsite. This process scrutinizes where these lubricants are utilized, how they are stored, handled, transferred, and dispensed to mitigate any potential contamination risks to lubricants and bearings. Undertaken by the lubricant manufacturer, this essential engineering endeavor seeks opportunities to streamline operations by standardizing the use of modern, often more cost-effective products across the site. The goal is to replace any redundant, unsafe, or outdated oils and greases while minimizing the variety of lubricants required for safe and efficient plant operations. Following the consolidation process, the number of lubricants stocked and utilized in many facilities can be reduced to less than half of the original count.

To kickstart the standardization process, the consolidation effort must identify all possible lubricant compatibility issues and propose appropriate engineered lubricant change-out and flushing procedures. Once a finalized list of new lubricants is established, the plant must embark on a series of steps to formalize the program:

1. Compile a formal list of approved lubricants for purchasing department personnel and establish a blanket purchase order for these products.
2. Notify all stakeholders of impending changes to the approved lubricant list.
3. Eliminate all non-approved lubricant stock from the plant.
4. Establish a protocol for stock rotation and control for all approved lubricants.
5. Acquire updated safety data sheets (SDSs) for all approved lubricants and discard non-approved SDSs (these include SDSs in the corporate emergency response plan (ERP)).
6. Procure dedicated, color-coded storage and transfer equipment for approved lubricants.
7. Label all approved lubricant reservoirs.
8. Replace all lubrication filters.
9. Develop a comprehensive lubricant change-out flushing procedure and systematically replace all non-approved lubricants in machine reservoirs, relabeling as necessary.
10. Update lubricant inventory control software with pertinent details such as specifications, suppliers, manufacturers, code numbers, min/max levels, and inventory turnover rates.
11. Revise affected preventive maintenance (PM) job tasks in the computerized maintenance management system (CMMS) to reflect new lubricant changes.
12. Adjust PM schedules in the CMMS as required.
13. Update equipment manuals to reflect changes in lubricants.
14. Update bill of materials (BOMs) in the PM work order system.
15. Amend the lubricant disposal procedure as necessary.

16. Adjust reporting requirements in the CMMS to reflect changes.
17. Conduct staff training to increase awareness of changes, product handling, safety protocols, and product disposal procedures.
18. Inform production staff of the changes.
19. Establish a procedure for conducting trials of new lubricants and gaining approval for their use.

4.10.1 New lubricant trial process

Following the implementation of a consolidation program, only approved lubricants may be introduced into the plant for regular use. However, the policy does not preclude the introduction of new lubricants on a trial basis. Should a trial be deemed necessary, a formal request must be submitted to the reliability/maintenance group via a designated "lubricant trial request form." This group will oversee the trial, ensuring that it adheres to established protocols.

A well-constructed trial request form should provide sufficient information to facilitate the trial and gather pertinent data for making an informed approval decision upon its completion. It should address the W5 questions—who, what, when, where, why, and how—and document the test results across seven distinct sections:

1. Who? Identify the trial requestor, lubricant supplier and manufacturer, and personnel involved in the trial.
2. What? Specify the trial lubricant's characteristics, including name, type, viscosity, additives, safety data sheets (SDSs), expected compatibility issues, and other relevant details.
3. When? Define the expected trial duration, commencement, and completion dates.
4. Where? Indicate the equipment on which the lubricant will be tested.
5. Why? Explain the rationale behind the trial, anticipated benefits, and expected outcomes.
6. How? Detail the test procedure, including disposal methods and testing conditions.
7. Results? Provide findings, conclusions, before-and-after data, and recommendations for approval or rejection based on test results.

TIP √ It is imperative to communicate the ongoing lubricant trial to plant personnel, employing signage to notify operators and maintenance staff of equipment undergoing testing. Before initiating any lubricant trial, consultation with approved lubricant manufacturers is recommended to ascertain compatibility and explore potential alternative products.

5

Contamination Avoidance and Control Strategies and Tips

5.1 Contamination Avoidance and Control Strategies and Tips One: Choose a Contamination Control Reservoir Breather

The act of breathing, allowing air to flow freely in and out of a closed space, is a fundamental function often taken for granted by humans.

In the realm of machinery design, engineers must be acutely aware of enclosed spaces and mechanisms that could lead to internal air pressure build-up. It is imperative that any machine or system design include provisions for relieving or ventilating excess air pressure at a controlled rate to restore or maintain a neutral or positively pressurized state within the space.

The capacity of internal mechanisms to "breathe" and equalize pressure plays a significant role in a machine's efficiency and the lifespan of its components. A classic example of this principle is evident in early combustion-engine designs, where rudimentary piston-and-ring technology often allowed combustion gases to seep past the piston rings into the crankcase—a phenomenon known as "blow-by."

Without engineered means for venting the enclosed crankcase, pressure would accumulate, leading to compromised seals and gaskets. This, in turn, would result in engine power losses, oil contamination, and persistent oil leaks. The solution to this problem came with advancements in piston and ring design and the implementation of a crankcase ventilation system in which fresh air is introduced through the filler cap, mingling with combustion gases before being expelled through the road-draft tube connected to the crankcase. This system evolved further with the introduction of the pressurized crankcase ventilation (PCV) system; a feature commonly found in modern engines to improve breathing and the environment.

The automotive crankcase serves as a reservoir akin to those found in gearboxes or hydraulic systems. In all instances, the reservoir functions as an enclosed container for housing oil pumped throughout the machine to lubricate bearing surfaces, ultimately returning to the reservoir within a closed-loop system. Properly designed reservoirs include an airspace, or headspace, above the oil level to accommodate thermal expansion and facilitate the de-aeration of the fluid, as aerated fluids can lead to pump cavitation.

To mitigate the pressure buildup resulting from fluctuations in oil level as the machine transitions from idle to full operation, and vice versa, a mechanism known as a breather is employed. However, it's crucial to note that while breathers allow for the free movement of air in and out of the reservoir, they also facilitate the exchange of any substances present within that airflow, including airborne contaminants and moisture—both of which can adversely affect the oil and the bearing surfaces it is intended to protect.

This necessitates a thoughtful approach by reliability engineers and maintainers in selecting the appropriate breather style and type for the prevailing conditions. Additionally, diligence in conducting preventive maintenance checks ensures that breathers remain in place, clean, and unobstructed, allowing them to fulfill their intended function effectively.

Breathers come in various configurations and styles (Figure 5.1), each tailored to accommodate different airflow, particulate sizes, and environmental conditions. Most breathers are consumable devices and should be replaced regularly as part of a comprehensive lubrication system maintenance program. The replacement schedule is determined based on the specific application and ambient conditions. In environments characterized by significant shifts in conditions or high humidity, breather caps equipped with filters and pressure valves offer added benefits, limiting air exchange and maintaining a positive suction head at the pump inlet.

Desiccant breathers represent a recent innovation in breather technology, offering enhanced air exchange and condition control. These units feature a transparent polycarbonate body filled with a silica-gel absorbent capable of retaining up to 40% of its weight in absorbed moisture. Visual media color change alerts operators to the need for replacement as the absorbent media reaches saturation.

Figure 5.1: Assortment of "tell-tale breathers.

Courtesy: Des-Case Corporation

TIP √ It's important to emphasize that a breather can only fulfill its function when properly installed. Breathers removed for reservoir filling or inspection purposes must be promptly replaced to ensure continued protection against external contamination. They constitute a critical component of any reservoir-based lubrication system and require their own dedicated maintenance schedule. Always write on the work order an action item to not only remove the breather but to also reinstall the breather!

5.2 Contamination Avoidance and Control Strategies and Tips Two: Understanding your Recirculating Oil System Needs

In various industrial settings, such as those in pulp and paper, petrochemical, energy, and steel industries, the operational machinery heavily relies on large-scale recirculating-oil lubrication and hydraulic systems. These systems are engineered to handle substantial oil volumes, often ranging from hundreds to thousands of gallons, to sustain continuous operations efficiently.

Recirculating-oil systems function by continuously delivering lubricating oil to numerous lubrication points throughout the equipment. This oil flows across bearing points and returns to a central reservoir for decontamination, cooling, and subsequent redistribution—a cycle that can persist round the clock depending on operational demands.

Operating in environments that are often open to air and prone to various contaminants, these systems face challenges in maintaining oil cleanliness. Contaminants such as production materials, moisture, and particulates can infiltrate the lubricant, posing threats to bearing surfaces and compromising the oil's effectiveness. If left unchecked, contamination can lead to abrasive wear on bearings, depletion of the oil's additive package, oxidation, and overall degradation of lubricating properties.

Traditionally, small to mid-sized recirculating systems manage contamination by implementing regular or accelerated oil change intervals based on cleanliness assessments. However, for large-scale hydraulic and recirculating lubrication systems, the downtime required for a single oil change can significantly impact asset utilization. Moreover, the costs associated with purchasing new oil, storing it, pre-filtering to meet cleanliness standards, and responsibly disposing of used oil can easily reach tens of thousands of dollars per change, not to mention the costs incurred from downtime.

To mitigate these challenges and reduce the frequency of oil changes in large-scale recirculating systems, four major "in-service" decontamination processes have been developed. These processes aim to address maintenance, reliability, and excessive downtime concerns:

5.2.1 Continuous full flow decontamination

This system involves pumping the entire oil volume from the reservoir through a strainer to remove large solid contaminants. The oil then passes through a primary filter element, typically a bag-style filter, to capture smaller contaminants before being cooled and delivered to machine elements or bearings. Duplexed filters are common in 24/7 industries, allowing for seamless switching between primary and secondary filters.

5.2.2 Continuous auxiliary decontamination

In scenarios where space constraints prohibit inline full-flow decontamination systems, an independent pump draws oil from the reservoir and passes it through various decontamination processes. These may include centrifuges for water extraction, large-micron filters for coarse contaminants, and small-micron depth filters for finer contaminants. The oil is then reintegrated into the main recirculation system after filtration.

5.2.3 Continuous bypass decontamination

Suitable for environments with controlled air filtration and temperature, this system type involves diverting a portion of the main oil flow through a continuous bypass loop. The diverted oil is filtered similar to full-flow systems, while the remainder continues through a cooler before reaching the bearings. This method minimizes contamination risk in filtered environments.

5.2.4 Periodic batch decontamination

In situations where no permanent filtration circuit exists, or for machinery with smaller footprints, a periodic batch-decontamination process can be employed. This involves shutting down the machine and evacuating the oil to a separate tank for decontamination either onsite or offsite by a third-party contractor. By reusing cleaned oil from other machines, this method significantly reduces oil-change costs and prolongs oil lifespan.

TIP √ Treat oil as an asset rather than a consumable. This simple thinking strategy can result in substantial financial benefits over the machine's lifecycle and contribute positively to the corporate bottom line. By implementing effective decontamination strategies, industries can enhance equipment reliability, minimize downtime, and optimize operational efficiency.

5.3 Contamination Avoidance and Control Strategies and Tips Three: When to Perform Contamination Avoidance (CA) versus Contamination Control

It is surprising how many companies still consider bearing failure and its associated costs as an acceptable part of business operations! This mindset is particularly prevalent among corporations operating in severe and semi-severe conditions where water, heat, and fine particulate matter are present.

Original equipment manufacturer (OEM) machinery typically includes built-in filtration elements as part of its fluid management systems, such as lubrication, hydraulic, and pneumatic systems. These elements act as sacrificial filtration devices, aiming to manage and control potential contamination issues to protect delicate bearing surfaces. They require monitoring, cleaning, or replacement when their media is close to being exhausted.

5.4 Understanding Contamination Avoidance and Control

Effective contamination avoidance and control are achievable with minimal effort when the requirements and basic relationships between machine, operator, and maintainer are understood.

5.4.1 Contamination control (CC) systems

Precision bearing surfaces rely on a lubrication film free of particles or water for protection. Filters trap and control contaminants before they can enter the lubricated zones and cause damage. Filters use passive surface attractant media to capture contaminants as the lubricant flows through or across the filter media.

Grease systems typically utilize heavy-gauge coiled wire mesh filters to capture large solid contaminants. Enclosed gearboxes and reservoirs employ breathers to equalize pressure and control solid and moisture contamination, with newer designs using desiccant media to prevent airborne particulates and moisture from entering.

5.4.2 Contamination avoidance (CA)

Contamination avoidance is the primary strategy for preventing premature bearing failure but often receives less attention in lubrication programs. This approach requires little capital outlay and fits seamlessly into any maintenance program, involving cooperation between operators and maintainers. Key elements of a CA program include:

1. Good housekeeping: Maintaining cleanliness and order on machinery and in lubricant storage areas.
2. Lubrication training: Educating personnel on the consequences of contamination and enforcing consistent processes and procedures.
3. Lubricant storage and transfer engineering: Using dedicated, color-coded storage and transfer equipment to protect lubricants from contamination.
4. Condition-based oil changes: Performing oil changes based on condition checks to avoid both excessive and insufficient filtration.
5. Lubricant cleanliness: Testing new lubricants and bulk fluids for cleanliness and additive package formulation before use.

Additional equipment and workspace changes may include implementing room ventilation systems and modifying machine designs to deflect contaminants away from bearing and lube reservoir areas.

TIP √ A proactive contamination avoidance program can significantly reduce reliance on contamination control measures and generate substantial cost savings. Try to make this your first approach toward decontamination.

6

Predictive Testing Strategies and Tips

6.1 Predictive Testing Strategies and Tips One: Employ a PCDA Strategy to Eliminate Wasteful Reactive Practices from your Lubrication Program

Predictive maintenance (PdM) has been a cornerstone of condition-based maintenance for over 50 years, rooted in the PDCA (plan, do, check, act) cycle first introduced in the 1920s as the Shewhart cycle by Walter A. Shewhart, and later renamed the Deming cycle in the 1950s by the Japanese in deference to W. Deming.

The PDCA process involves understanding how systems/components deteriorate, using suitable predictive technology/methods to detect current conditions, analyzing trends, and taking corrective action when necessary.

When implementing a proactive lubrication program, following these few rules can ensure success:

Follow the PDCA cycle: This involves planning, executing, checking results, and taking action based on those results.

Provide operator training: Proper training ensures accurate data collection and interpretation, vital for effective decision-making.

Maintain data collection accuracy and consistency: Consistent data collection under the same conditions is crucial for developing accurate trends.

Act immediately on predicted failures: When imminent failure is detected, swift action is essential to minimize downtime.

The theory behind successful methodology is straightforward: Collect data consistently, analyze trends, and take action when control limits are breached.

A structured approach to implementation of a PDCA approach involves a ramp-up strategy:

Choose a suitable pilot: Select a pilot asset or area with a history of failures or easy access for testing PdM methods.

Select a predictive method/tool: Choose a method/tool based on immediate need, budget, or legacy. Options include fault-code analysis (FCA), wear-particle analysis (WPA), or infrared thermography (I/R).

Act on findings immediately: When impending failure is detected, create work orders for planned repairs linked to the PdM results to ensure action is taken promptly.

Incrementally rolling out the program into other areas based on the success of the pilot ensures a manageable transition from reactive to predictive maintenance. With proper planning, training, and execution, the shift to predictive maintenance is achievable and offers significant benefits in terms of reducing downtime and maintenance costs.

TIP √ A suitable pilot asset for your initial effort can include equipment in a designated area, a group of similar equipment, a manufacturing line, or a specific component.

TIP √ When sampling oil for analysis, always pre-flush sampling equipment and draw your samples from the same spot, using the same procedure and equipment each time.

6.2 Predictive Testing Strategies and Tips Two: Implement an Industrial Oil Analysis Program

In implementing an industrial oil analysis program, you will be putting in place a mature predictive methodology employed with incredible success around the world with many world-class laboratories to choose from as your oil analysis partner. A successful program involves several steps to ensure its success and effectiveness in optimizing machine health and lubricant condition. Here are the key eight steps to follow:

1. Appoint a program champion: Select a supervisor or manager level individual who will advocate for and oversee the implementation of the oil analysis program.

2. Choose a suitable pilot area/machine: Select a critical piece of equipment or area for the initial implementation of the program based on factors such as criticality, downtime costs, and product quality.

3. Conduct a lubricant audit: Identify and verify the lubricants currently in use at the plant by checking work order systems, lubricant identification labels, and safety data sheets (SDS). Ensure that the correct lubricants are specified for each application.

4. Choose a laboratory: Select a reputable oil analysis laboratory that specializes in industrial sample testing and can provide timely and consistent results. Consider factors such as sample turnaround time, testing consistency, and long-term relationship potential.

5. Set up the pilot sampling program: Work with the chosen laboratory to establish a sampling program that includes acquiring necessary sampling hardware, extraction pumps, and sample bottles. Train staff to take clean oil samples that best represent the condition of the lubricant and its particulate levels.

6. Take virgin oil samples: Collect and send off virgin oil samples of all lubricants to be checked in the pilot program. These samples will serve as reference benchmarks for the laboratory to compare against operating samples and identify any abnormal conditions.

7. Implement regular sampling: Set up a schedule for regular oil sampling from the pilot area/machine, ensuring that samples are taken from live fluid zones where the lubricant is freely flowing. Sample points should be located downstream of lubricated areas to capture wear elements before they are filtered out.

8. Analyze results and take action: Review oil analysis reports from the laboratory and take appropriate action based on the findings. This may include adjusting oil change intervals, identifying incipient bearing failures, or diagnosing root causes of failure.

By following these steps and incorporating oil analysis into your maintenance strategy, you can optimize machine reliability, reduce downtime, and improve overall production throughput in your industrial plant.

TIP √ Consider combining your predictive strategy with a plan-do-check-act approach as part of your oil analysis program.

6.3 Predictive Testing Strategies and Tips Three: Getting World Class Results from your Oil Analysis Program Requires Using the Right Container to Collect your Oil Sample

Used-oil analysis is a highly respected, effective, and inexpensive method employed to determine when to change oil based on its condition, to predict incipient bearing failure so that relevant action can be taken in a timely manner, and to help diagnose bearing failure should it occur.

Although laboratories offer and perform similar testing capability on similar equipment, they will all differ slightly in their processes, methods, and reporting capabilities. Success is therefore primarily driven by methodologies that use consistent procedures to collect, test, and report on every sample.

Having an established relationship with your laboratory allows them to understand and compensate for working and ambient conditions for each machine being sampled. This will make it possible to develop reliable trending for each sample point.

Arguably, the most important and misunderstood element of the used-oil-analysis process is the consistent collection of a quality representative oil sample for testing purposes. To achieve this, the maintainer must ensure that the sample is collected on an established frequency and in an appropriate manner to maximize data density and minimize data disturbance. This requires the choice and consistent use of appropriate sample containers of the right material and size.

6.3.1 Containers and sample results

If an oil sample not exactly representative of the oil system being tested, the result will be compromised—the test is only as good as the sample! Once a sample is extracted, it should be placed into a clean neutral environment that will not adversely affect the sample integrity as it is shipped from the machine to the laboratory for testing. In addition, the amount of fluid collected for testing should meet laboratory requirements.

TIP √ Ask your oil analysis laboratory what amount of lubricant they require to adequately perform their laboratory tests.

To truly understand your laboratory's needs, take a tour of their oil-analysis facility and you will be surprised at the variety of containers used to collect and send used-oil samples to the laboratory. These can include used screw-top soda pop or milk containers, mason jars, plastic pill bottles, and ketchup and condiment containers, to name a few. It should appear obvious that any used-oil sample placed in a previously used container will likely have trace particulate and/or fluid contamination from its original fluid that is not representative of your oil system. To achieve a true representative sample, it is imperative to use a container specifically designed to hold oil samples. Oil-sample containers are sold according to their cleanliness level, material and size.

Table 6.1: Sample container cleanliness levels.

Clean designation	# Particles >10 microns/ml	ISO designation
Ultraclean	1	< ISO 9/7
Super clean	<10	ISO 10/8 to 11/9
Clean	11–100	ISO 12/10 to 14/11
Dirty	>100	>ISO 15/12

Courtesy: ENGTECH Industries Inc.

Sample containers are available in three basic cleanliness levels: (1) clean, (2) super clean, and (3) ultra clean. Cleanliness is defined by the residual particulate greater than 10 micron/ml of fluid that might be expected to contaminate the sample. Refer to Table 6.1 for particulate range and materials by cleanliness levels.

The appropriate cleanliness-level container will depend on the application and is based on the cleanliness levels required by the machine being lubricated. This is established using a signal-to-noise-ratio (SNR) calculation that divides the oil target cleanliness level (signal) by the bottle contamination level (noise). For example, if the target level of a fluid is 50 particles that are >10 microns/ml and a super-clean bottle is chosen, the SNR calculation would be 50/10 = 5. That signifies a respectable SNR value of 5. The higher the number, the cleaner the sample will be.

6.3.2 Container material and size

The three most common container materials are glass, high-density polypropylene (HDPE), and clear PET plastic. The glass and clear PET plastic containers offer great visual analysis capability for immediate detection of water and heavy or large contaminants. Glass is the most expensive option but offers the cleanest environment for storing the oil and is compatible with all oils. PET plastics start to deform and break down with fluids exceeding 200 F, but offer visual clarity for lower-temperature fluids and are less expensive than glass. HDPE is opaque and is not as clean as glass or PET, but is the least expensive option for less-critical types of fluid analysis. See Figure 6.1 below.

Figure 6.1: Three accepted oil analysis containers.

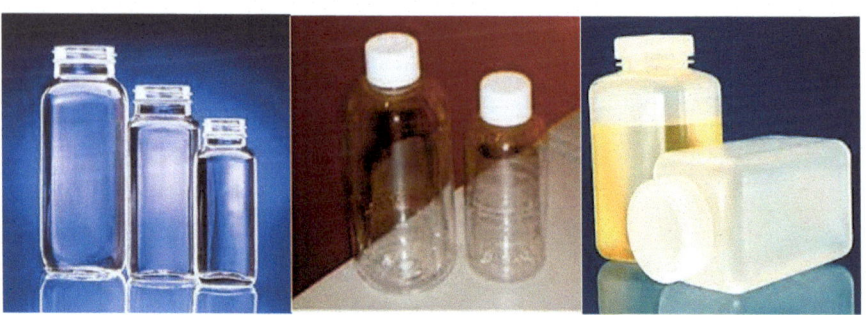

GLASS vs PET PLASTIC vs HDPE POLYETHYLENE

Courtesy: **ENGTECH** Industries Inc.

Where lubricant contamination is critical, such as in hydraulic-fluid testing, glass is always the best option.

Containers sample sizes vary from 2 oz. (50 ml) to 8 oz. (200 ml) in size. Size will depend on whether a single or multiple tests are to be conducted on the sample—your laboratory will assist you in your decision here. When filling the sample bottle, the amount of lubricant will vary based on oil viscosity and whether the laboratory needs to agitate the sample in the bottle. A rule of thumb is if the lubricant viscosity VG is lower than ISO VG 32, fill to 75%; ISO 32 to 100, fill to 66%; greater than ISO VG 100, fill to 50%.

When an oil-analysis program is in place with a dedicated laboratory, the laboratory will be able to advise the best type and size of sample bottles to use and their approximate fill-level requirements to achieve maximum value density (the most usable data from your sample), minimum data disturbance (cleanest representative sample), and most accurate reporting capability. In most cases the laboratory is able to provide the correct sample bottles or put you in touch with a recommended supplier.

TIP √ While glass containers are the most expensive, glass offers the cleanest environment for storing samples and is compatible with all oils. The time and effort to take a true representative sample from the machine is far more expensive that the cost difference between a glass container and its counterparts.

TIP √ Container size depends on the number of tests that will be conducted on the sample. Sample quantity varies based on oil viscosity and whether the sample needs to be agitated in the container. Your testing laboratory will help you determine the sample size required for your oil samples. If the lab allows you to use dirty containers, find another testing service.

7

Weather Related Strategies and Tips

7.1 Weather Related Strategies and Tips One: Use of Outdoor Lubrication Systems in Cold Climate Winters

There are many instances in which lubrication systems are employed outdoors. In geographical areas where the winter months are cold, lubricant viscosity will thicken, especially when a petroleum-based lubricant (non-synthetic) is used.

For automobiles, the solution has been provided by the oil companies with the adoption and use of multi-grade lubricants such 5W30, 10W40, 0W5, etc. The "W" designates the oil as winter ready. See Figure 7.1 for pour tests in cold weather. Therefore, a 5W30 viscosity motor oil is designed to act as a 5-weight viscosity oil (thin oil) when cold and thicken to a 30-weight viscosity oil when at operating temperature in the engine. This allows the vehicle to be more easily started, getting vital protective lubricant quickly on to the valve train as the engine turns over, with less drain on the battery.

In the world of industrial oils and greases, lubricants are mostly single weight viscosity that if heavy enough and cold enough will thicken to the point where they will stall any auto-lubrication system and potentially starve the bearing.

Sometimes this is difficult to test for in remote locations where single point auto lubricators (SPL) using #2 grease are employed. The cold grease will stiffen and the lubricator will stall until the temperature rises and it starts to work again. Meanwhile the bearing may not have been lubricated for months and will be in a rapid failure state when all appears normal. A similar scenario will

Figure 7.1: The effect of temperature on the pourability of different gear oils. Note: at −30°C the synthetic oil pours clear compared to its petroleum-based cousins.

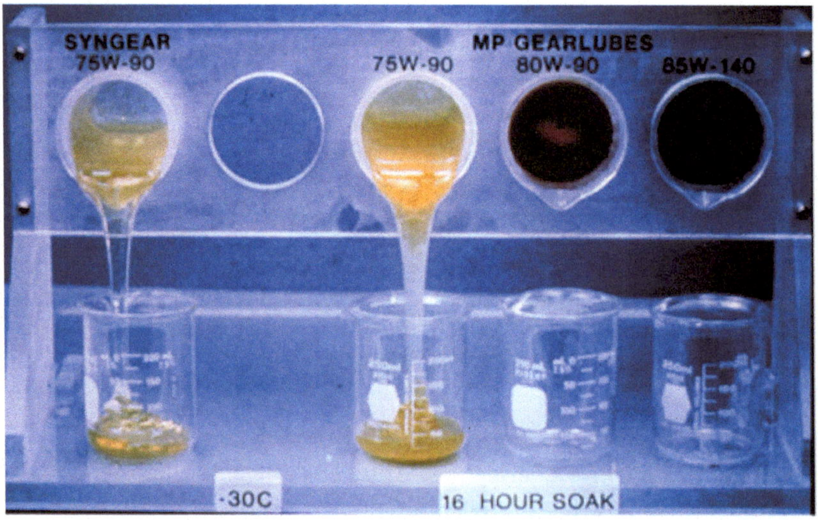

Courtesy: Petro-Canada

play out with reservoirs connected to multipoint lubrication delivery systems using higher viscosity lubricants. Knowing and expecting this to occur is the solution to continued lubrication in any temperature. The following tips will help put together a cold climate strategy for your outdoor lubricated bearings.

For lubricant delivery systems deployed in cold winter climates, the use of the correct lubricant viscosity is essential for continued lubricant delivery throughout the year. Therefore, if the manufacturer calls for a #2 bearing grease, this will need to be exchanged prior to the cold season for a less viscous (thinner) grease—namely, a #1 or #0 grease depending on the temperature drop. Once the cold season is over, the less viscous lubricant can then be changed

back to the original #2 grease specification.

TIP √ Consider inquiring with the lubricant supplier engineering department the possibility of permanently switching to the same lubricant in a #1 viscosity or lighter lubricant so no seasonal replacement is required.

TIP √ If lighter grease or oil viscosity cannot be used, a number of thermostatically activated heating devices can be employed that include:

- A 12v or 110cv battery warming blanket wrapped around the grease or oil reservoir.
- An oil dipstick heater.
- An oil pan type heater similar to an auto frost plug heater.
- Placing the lubricator pump and reservoir into a heated cabinet.
- Trace heating of the lube lines and reservoir.

8

Lubrication Safety Strategies and Tips

8.1 Lubrication Safety Strategies and Tips One: The Role of the Safety Data Sheet (SDS) in a Lubrication Management Program

Safety data sheets (SDSs) serve as vital documents for handling potentially hazardous substances such as lubricants. These sheets provide comprehensive information on the product's properties, hazards, and safety precautions, ensuring the well-being of those who work with these substances and safeguarding the environment in case of accidents.

The SDS is a legislated document required for every potentially harmful product sold across the world that meets the international requirements of the UN Globally Harmonized System (GHS) of classification and labeling of chemicals. The Occupational Safety and Health Administration (OSHA) is the US governing agency that legislates under the Hazard Communication regulation to ensure that all workplace employees coming into contact with potentially harmful substances—such as lubricants—have ready access to the current SDS. See Figure 8.1.

Comprising 16 sections, SDSs systematically detail crucial aspects of the product. Beginning with Section 1, identification, these sheets clearly define the product, its manufacturer, intended uses, and any limitations. Sections 2 to 8 cover hazard identification, composition, first aid measures, firefighting protocols, spill response procedures, and handling/storage guidelines. This information empowers individuals to handle lubricants safely and effectively.

Sections 9 to 16 delve into technical details, ecological considerations, disposal instructions, transportation guidelines, regulatory compliance, and

Figure 8.1: Shop floor hazard awareness area with SDSs.

Courtesy: **ENGTECH** Industries Inc.

additional information. While these sections may contain more specialized data, they remain essential for ensuring thorough understanding and adherence to safety protocols.

Proper training on interpreting and utilizing SDSs is imperative for personnel handling hazardous substances. Access to SDS manuals, whether in print or online, ensures that workers have immediate access to critical information necessary for safe practices.

In essence, adherence to SDS guidelines is paramount in preventing accidents, mitigating environmental impact, and upholding workplace safety standards within industries dealing with potentially harmful substances like lubricants.

TIP √ Safety data sheets are used to identify dangerous consequences before or when a spill or accident occurs. Refer to the SDS to develop and improve the corporate emergency response plan (ERP).

TIP √ Ensure SDS plans are well identified in all major congregated areas of the plant (see Figure 8.1).

8.2 Lubrication Safety Strategies and Tips Two: Making Safety Priority One!

When it comes to the handling and transfer of chemicals and lubricants, safety has to be priority one. The safety dress code for maintainers and lubrication technicians must be specific and clear. The need for the correct personal protective equipment (PPE) is absolute. Typical PPE shall minimally include protective footwear, coveralls, gloves and eyewear designed to protect against skin and eye contact exposure to oils, greases, and chemicals when handling, transferring, or cleaning up spills.

TIP √ As each plant or workplace supports the use of a unique inventory of lubrication fluids and chemicals, determining the correct PPE can be facilitated using a current list of all lubricants and chemicals on site. This list is used to assemble an up-to-date set of safety data sheets (SDSs) for each lubricating fluid and chemical used in the plant/workplace from the lubricating fluid supplier(s). The SDSs are then referenced to build a PPE requirement list for each fluid from the PPE requirements found in the SDS—Section 8: Exposure Control/Personal Protection. Note: for fluids that require additional non-standard PPE (e.g., use of a respirator for lubricants chemicals and solvents identified as potential hazardous to lungs) these are best identified and listed as an additional PPE requirement on the work order. This list and SDS library are now recognised as an asset and must be maintained like any other asset if it is to be current and effective.

Lubricating fluids can weigh many pounds/kilos. For example, a standard 55gallon drum of lubricating fluid can weigh almost 500 lb (225 kilo), making proper training on the correct use of lifting devices, slings, and ergo devices essential. In addition, all footwear should be grade one safety certified with steel toe and puncture proof, oil resistant soles.

8.2.1 Operating pressured lubrication devices safely

Many manual and automated lubrication delivery systems such as grease and oil systems, hydraulic systems, and manual pressurized grease gun systems operate at high pressures that require operators and machine lubricators to be aware of the dangers of high-pressure injection hazards that can prove lethal if unattended to.

Most people are blissfully unaware that simple electric/compressed air grease guns that can be purchased from any hardware store can deliver grease (and oil) at nozzle pressures up to and greater than 5000 psi. And, that a simple innocuous hand grease gun can develop even higher pressures up to 15,000 psi.

In a recent 2020 study published by Dr. Stewart O. Sanford titled "Management of High- Pressure Injection Injury of the Hand in the Emergency Department" [1] Sanford states, "high-pressure injection injuries occur when a high-pressure injection device such as a paint or grease gun injects materials into the operator [skin]. This injury most commonly occurs in the dominant hand and index finger". Sanford goes on to say "the injection typically occurs to the fingertip when the operator is trying to wipe clear a blocked nozzle, or to the palm of the hand when the operator is attempting to steady the gun with a free hand during the testing or operation of the [grease gun] equipment".

Similar injuries also occur when performing condition and leak checks on pressurized hoses and lines (especially flexible line material). It is not uncommon to see maintainers, and machine operators in a TPM (total productive maintenance) environment, physically running their hands over high-pressure flexible lines feeling for leaks or soft spots that can easily inject fluid into the hand if a leak is found. Leak checks are best performed using a piece of stiff card, NEVER your arm, hand or fingers.

Grease gun training instructs the operator never to place their finger directly over the flex delivery tube or nozzle tip when activating the trigger. If cleaning the nozzle tip, always take the non-dominant hand off the trigger or pump lever. Unfortunately, too many companies do not provide enough proper grease gun training to its maintainers, or it's machine operators.

Injection through the skin can occur under 1000 psi, and the higher the pressure, the more damage can occur to the structure of the hand or finger. If you have ever in the past received an inoculation in your arm through use of a medical pressurized jet injector, you will recall the "pressure" felt was fairly low as these devices delivered their vaccines somewhere between 1000 to 2000 psi. It is easy to ignore a pressure injection injury, second-guessing if it did or did not occur. Sanford goes on to state "a high-pressure injection injury should be considered a surgical emergency". In the Northwest Linemen College, 2018 YouTube safety video "Hydraulic Injection" [2] it is visually demonstrated how a seemingly small injection site showing a small area of redness can accelerate to a devastating injury within hours if not medically tended to immediately.

In the Sanford study, he found the overall incidence of amputation was 48% with solvents causing the greatest damage followed by grease injection. In 25% of all cases amputation occurred. Similarly, in an earlier medical study performed at the New York Methodist Hospital by Snarski and Birkhan, they found that high-pressure grease guns/systems were responsible for over 50% of all injection injuries, seconded by hydraulic fluid injuries. In addition, they found that the overall incidence of medical amputation resulting from such injuries was 48%, and closer to 100%, when the injection pressure was greater than 7000 psi.

Even if the limb is saved, the long-term outcome of injured persons is not always a good one. In Sanford's study he found that those who underwent successful surgical debridement, 50% ended up with a reduced range of motion; injuries adversely affected daily living activities; grip strength was reduced by 35% in 75% of cases; and all patients continued to suffer from some level of neuropathic pain.

If you suspect a high-pressure injury has occurred, you must seek professional medical treatment *immediately*. All studies show that the quicker the injury is recognized and professionally tended to, the higher the chance to limit the damage and save the limb.

TIP √ Should you feel an injection injury has occurred, list all of the following Sandford recommendations that can allow the attending physician to determine the severity [and treatment] of the injury;

- Type and viscosity of the material injected (a copy of the fluid type SDS is essential here).
- Time interval between injury and treatment (log the time the injury occurred).
- Amount of material injected and velocity of injectant (if unknown take the lubricating device with you).
- Pressure of the appliance (or system).
- Anatomy and distensibility (swelling amount) of the site of injection.

TIP √ Whenever possible, it is always good to use a cell phone to photograph the injury sustained to the limb as soon as possible after the injury, along with a photo of the injection device.

In their safety video, the Northwest Lineman College recommends the following first response plan:

- Treat immediately (seek medical attention)
- Don't let victim drive
- Don't leave [victim] alone

- No food or drink
- Immobilize and elevate the wound.

If you feel you have ingested lubrication fluid, have skin or eye irritations through exposure, or believe you have suffered a pressure injection injury, seek medical advice *immediately*.

Bibliography

[1] Sanford, S.O., Mills, T.J. (Ed.). (2020, March 20). *Management of High Pressure Injection Injury of the Hand in the Emergency Department.* Medscape. https://emedicine.medscape.com/article/826620. (Accessed March 5, 2021)

[2] Northwest Lineman College. (2018). (*Hydraulic Injection*). Available at: https://www.youtube.com/watch?v=O9n23cY65bc. (Accessed March 5, 2021)

Index